Ion Channel Kinetics

Rhodri J. Walters

Copyright © 2017 Rhodri J. Walters PhD
All rights reserved.
ISBN: 1978432658
ISBN-13: 978-1978432659

DEDICATION

This book is dedicated to my many teachers over the years, notably Professor Tim Jacob for his introduction to the world of ion channels; Dr Keith Collard for imparting the wonderment of synaptic signalling and second messengers; Professor Francisco Sepulveda for providing the opportunity to become an electrophysiologist; Professors Len Stephens & Phillip Hawkins of the University of Cambridge for demonstrating what it means to be elite scientists and for their introduction to molecular biology; and to Dr Jahanshah Amin of the University of South Florida, one of the world's great molecular biologists, for teaching me site-directed mutagenesis. This book is thus dedicated to all my mentors. I would also like to thank those other great scientists whom I have had the privilege to meet and to be inspired by, notably Professors David Colquhoun, Sir Bernard Katz, Richard Kramer, Heinz Wässle, Robin Irvine, Michael Berridge, Eric Bennett, Thaddeus Bargiello, Vitas Verselis, Robin Lester, Peter McNaughton & David Sattelle, *inter alia*.

Table of Contents

DEDICATION iii
FOREWORD viii
ACKNOWLEDGEMENTS ix

Chapter One 1
The nature of electricity in living systems 1
The intimate relationship between current, voltage and resistance 2
Electrical power and its loss 3
A brief introduction to membrane capacitance 6
Recording from cell membranes with electrodes 7

Chapter Two 10
Probing an electrical black box 10
The biophysical question 11
The vectorial transport of ions across polarised epithelial cells 11
Investigating the electrical properties of the crypt using the patch-clamp technique 13
Effects of ion substitution experiments upon crypt membrane potential 14
The Nernst equation 16
The Goldman-Hodgkin-Katz equation 17
Whole cell currents recorded from unstimulated small intestinal crypts 18

Chapter Three 21
Exciting the membrane 21
The effect of carbachol upon crypt membrane conductance 21
The effect of VIP upon crypt membrane conductance 23
Chapter Four 30
The enigma of E_{Cl} 30
Is GABA a bipolar neurotransmitter? 32

Chapter Four 34
In search of the underlying single channels 34
Calculating single channel conductance 36
An introduction to single channel kinetics 39
Noise and filtering considerations 40
A second single channel activity within the crypt basolateral membrane 44
Engineering ionic gradients in single channel recording 50

Chapter Five 55

Recording the activity of very small ion channels 55

Chapter Six 62

The anatomy of action potentials 62

The anatomy of an action potential 62

How fast are action potentials? 64

How do we record action potentials? 65

How can we distinguish between excitable Na+ and Ca2+ channels? 65

Do dendrites contain excitable ion channels? 66

The challenges of recording currents from neurons 67

Chapter Seven 71

The kinetics of action potentials 71

Using incremental voltage steps to separate voltage-gated ion channels 73

Glucose-activated current deflections exhibit strongly voltage-dependent kinetics 74

Establishing the ionic selectivity of the inward deflections 77

Effects of Ca2+ channel inhibitors and chelators upon the recurrent deflections 80

Kinetic properties of the high threshold inward current 82

Voltage-dependent properties of the G-type current 83

Voltage-dependence of the glucose-activated Ca2+ channels in isolated β cells 86

    Space clamp considerations 86

    Calculating the coupling coefficient 87

Chapter Eight 91

The kinetics of ligand gated ion channels 91

Amino acid substitution at position S439 produces variable channel kinetics 93

Substitution at 439 produces only small changes in apparent affinity 97

Effects of amino acid substitution at S439 upon mean current amplitudes 99

Substitution at adjacent residues has little influence upon channel kinetics 99

The kinetics of channel deactivation change with agonist concentration 102

Fast perfusion reveals a time-dependence in the number of open states occupied 103

Correlations between residue properties and measured kinetic parameters 108

Conclusions 108

Chapter Nine 112

Final reflections 112

ABOUT THE AUTHOR 114

APPENDIX A 115

Whole cell recording methods for small intestinal crypts 115

APPENDIX B 119

Single channel recording methods for small intestinal crypts ... 119
APPENDIX C ... 121
Limitations of the voltage clamp technique ... 121
APPENDIX D ... 123
Expression of $GABA_A$ receptor subunits in PC12 cells ... 123
APPENDIX E ... 129
Data for excised patch ion substitution experiments ... 129
APPENDIX F ... 132
Experimental methods for recording from β-cell clusters ... 132
Cell culture ... 132
Solutions ... 133
Analysis ... 135
Current clamp recordings ... 137
Calcium imaging ... 137
Single-cell recording and selection ... 137
Limitations imposed by inability to perform capacitance measurements ... 138
APPENDIX G ... 140
Recording GABAergic currents from Xenopus oocytes ... 140
Methods ... 140
    Oocyte preparation ... 140
Electrophysiology ... 140
Perfusion systems ... 141
Analysis ... 141
Exponential fitting ... 142
APPENDIX H ... 143
Studying ion channels through cell volume regulation ... 143
APPENDIX I ... 154
Cystic Fibrosis: the search for an alternative chloride channel ... 154
APPENDIX J ... 168
Schizophrenia: A cyclical and heterogeneous dysfunction of cognitive and sensory processing?
... 168
APPENDIX K ... 210
Seeing a change against the light: how neural circuits are adapted in the retina ... 210
APPENDIX L ... 229
Further reflections on the role and function of GABA receptors ... 229
$GABA_C$ receptors ... 229

GABA$_A$ receptors 231
GABA – the bipolar transmitter 232
APPENDIX M..239
Bibliography 239

FOREWORD

Scientists have tended to approach the study of electricity, or excitability, within biological systems empirically and, by extension, biology has therefore fragmented into myriad domains, each with its own lore and wisdom. As electrophysiology is a province within the realm of biophysics, the subject has traditionally been approached from a mathematical standpoint, often at the expense of a more holistic approach to understanding the nature of biological systems. To this extent, we need to be careful that our theoretical abstraction from the biological context, and our use of tools which are invariably prone to artifacts, do not disrupt or distort the very essence of what we are seeking to understand. We should, therefore, avoid the pitfalls of failing to understand the nature of living systems and of falling prey to the absolutism of physics, a paradigm of scientific thought which tends to refute measurements or findings simply because we cannot explain them theoretically or model them mathematically.

The field of biophysics encompasses the disciplines of biochemistry, electronics, optics, physics, mathematics, physiology, genetics, physiology, and pharmacology. It is, therefore, an advanced field, one which is seldom encountered as an undergraduate. This should not, however, preclude us from attempting to explain the fundamentals of electrophysiology in broader, more conceptual terms, rather than spinning an opaque web of mathematical intricacy and complexity. It is in this spirit that we shall introduce the intriguing world of ion channels, avoiding a rigid use of mathematics and religiously deriving innumerable equations from first principles *(as this has been done already)*. Throughout this work, we shall introduce key concepts progressively, crossing each bridge in turn as we come to it.

Yet Einstein's laws of relativity and Newton's laws of motion are as beautiful as they are penetrating. The insights they provide are simple, yet powerful. Within a few symbols, they impart a greater understanding of the natural world than could be gleaned from a thousand words of descriptive narrative. Thus, an equation serves as a wondrous dynamic equilibrium whose symbols come to predictive life. This book will seek to approach an understanding of ion channels in the same vein, using those practical problems which I have been privileged to solve from first principles. However, the use of mathematics throughout this work has been sparing, as we cannot hope to understand the living, breathing cycles and patterns of nature through cold-minded abstraction and reductionism. Finally, these biophysical solutions are then placed into their biomedical context within the appendices so that their greater significance is not lost.

ACKNOWLEDGEMENTS

I would like to thank all those who, over the years, have allowed the author to unleash his curiosity within their laboratories and to explore his original thoughts and ideas. This is an especially rare privilege within the modern laboratory, a closed world in which grant revenues are measured in terms of 'dollars per square foot' of laboratory space. Thus, there is constant, unrelenting pressure to produce and publish in accordance with a research agenda that has previously been prescribed and funded within a grant 'contract'. Naturally, without such a license to explore, much of the experimental work presented here would never have been made possible. As the majority of the work contained within this book has previously been presented at international seminars and conferences rather than formally published in academic journals, this compendium of research into the properties and nature of a diverse set of ion channels affords an opportunity to pass the flame of wonderment on to a new generation of scientists. For those whose curiosity remains unsated, additional musings are published within the appendices.

Chapter One

The nature of electricity in living systems

At the climax of the 19th Century, the longstanding feud between Tesla and Edison reached its climax in a heated debate over the relative merits and drawbacks of transmitting electricity by means of alternating or direct current, respectively. At stake was the contract to illuminate the 1893 World's Fair in Chicago *(which in of itself was worth a King's ransom)* and inestimable scientific *kudos*.

Edison argued that the high voltages employed by Tesla's alternating current system were dangerous and even went to the extreme of publicly electrocuting an elephant to illustrate his case. Tesla, undaunted, duly harnessed the power of Niagara Falls and soon after Chicago became a festival of light.

The challenge that Tesla and Edison faced was that the transmission of electrical energy over long distances incurs a loss of power, and there are only so many electrical substations that can be built along the way. A not altogether dissimilar dilemma is encountered within biological systems. How, for instance, does a threatened giraffe send a swift series of electrical impulses from its elevated brain to its leg muscles without any substantial attenuation of the signal in a fight-or-flight situation? In other words, how does the central nervous system survey its environment, make a calculated decision, and send a measured signal over six metres in less than a second.

In pondering these two, not entirely unrelated questions, we encounter two fundamental laws of electricity which apply to biological systems just as much as they do to a power grid, albeit on a very different scale. The first is the relationship between power (P), voltage (V; or *electromotive force*), current (I), and the resistance to its flow (R).

First and foremost, we shall consider a fundamental difference between biological and engineering contexts. Within biological systems, the flow of current is usually mediated by both positively charged ions, known as cations *(as they flow to the negative terminus of a battery, or cathode)* and negatively charged ions, known as anions *(which flow to the positive terminus, or anode)*. The most commonly encountered cations within biological systems are $Na+$, $K+$, and $Ca2+$ (although other important cations exist, including free protons - *i.e.* $H+$ ions - and $Mg2+$), whereas the most frequently encountered anions are Cl- and HCO_3^-. Fortunately, ion channels tend to simplify matters greatly, given that they are preferentially selective to one specific *species* of anion or cation, and are therefore classified as being Na+ channels, Cl- channels, Ca2+ channels, and so forth.[1] *In most solid-state electrical systems,* however, we talk only in terms of the transmission of electrons flowing from a negative polarity, or terminal, towards a positive one. Thus, within biological systems, currents are mediated by the flow of positive and negative charges across potential gradients, whereas in electrical systems we tend to think only in terms of the flow of electrons.

However, most of the physical principles governing electricity remain sacrosanct, despite the subtlety of their application in biological systems. Before venturing further, we should perhaps briefly recapitulate the fundamental properties of electrical circuits before we look at the intricacies of biological ones.

The intimate relationship between current, voltage and resistance

If the voltage (V) is simply viewed as an electromotive force, then the current (I) that flows between two points of divergent electrical potential (V) is given by the difference in potential divided by the electrical resistance (R) encountered between the two points. This is otherwise known as Ohm's Law, which is simply expressed in the following form;

$$V = I.R$$

It should be noted that biological systems tend to operate with currents and voltages which are many orders of magnitude smaller than those encountered within physical systems. Whereas Tesla created coils which could generate vast voltages of over a million Volt(s); and electrical power lines typically transmit 'small' alternating currents of hundreds of Amperes at very high voltages (115 to 1,200 kV) over vast distances at 50 Hz; even biological action potentials, which are transmitted as alternating waves of polarizing current are typically limited to +/- 80 mV at frequencies of up to 100Hz in the neocortex.[2] Although currents of many hundreds of Amperes (kA) are routinely conducted by power

---

[1] Please note that perfect selectivity is an assumption as most ion channels are permeable to more than one species of ion.
[2] de Kock CP, Sakmann B. High frequency action potential bursts (>100 Hz) in L2/3 and L5B thick tufted neurons in anaesthetized and awake rat primary somatosensory cortex. J Physiol. 2008 Jul 15;586(14):3353-64. doi: 10.1113/jphysiol.2008.155580. Epub 2008 May 15.

lines, within biological systems we commonly measure currents which are in the range of thousandths (mA) or millionths of an Ampere (μA) within tissues; thousandths of a millionth of an Ampere across cell membranes (nA); and as small as a millionth of a millionth of an Ampere for individual ion channels (pA).

In electrical systems Ohm's Law is seldom obeyed for the simple reason that currents flowing through wires generate not only electrical fields *(which we shall not consider these further here, even though these are what we measure when we use surface electrodes to study the output of the brain or heart)*, but also **heat**. Heat is that electrical energy, or power, which is dissipated as current flows through a resistor, or conductor,[3] and conductors usually exhibit a change in their resistance as they experience a change in their temperature. Thus, true Ohmic conductors are seldom encountered within electrical circuits and are rarely observed within biological systems due to the constantly changing properties of individual ion channels.

Electrical power and its loss

Within biological systems, a great deal of the cell's energy is expended in the generation and replenishment of electrical potential gradients across its various membranes. These are dissipated every time ion channels open, for instance when an action potential 'fires'. In living cells, this dissipation of electrical energy in the form of an ionic current is used to drive such diverse processes as muscle contraction, nerve conduction, or the electrochemical generation of ATP.

Within physical systems, power (P) is defined as the rate of performing work or, in electrical terms, the rate at which an electrical charge measured in Coulombs (Q) is driven across an electrical potential difference (our old friend V again). However, as the rate of movement of charge (Q) with time (t) is what we term current (I), electrical power (P) is simply calculated as the product of the current (I) and voltage (V). This can be presented in the following form:

$$P = V.I$$

---

[3] Electrical conductance (G) is the inverse of resistance (R), namely $R=1/G$

However, this really doesn't tell us a great deal about the dissipation or 'loss' of power. In a high voltage cable, power loss (P) is given by the product of the resistance R and the square of the current (I), which quickly explains why direct current (DC) failed to illuminate the Chicago World's Fair, as larger currents at lower transmission voltages (DC) exhibit a greater power loss than smaller alternating currents at higher voltages (AC). This is summarised by the following equation;

$$P = I^2.R$$

In other words, if you pass a significant current through a substantial resistance, you are going to generate a great deal of heat and, as Edison famously observed when passing a direct current through a tungsten filament, light. While generating light may be a useful product of the dissipation of electrical energy, and heat is not always unwelcome during the long winter months, losing electrical energy in the form of heat across long-distance power lines is a generally undesirable outcome.

Perhaps surprisingly, there are biological correlates, for example, the generation of heat within the mitochondria of brown adipose tissue, but such instances are rare exceptions to the rule that biological systems have evolved to be very efficient when it comes to electrical energy. As currents measured within biological systems are typically only in the range of millionths or billionths of an Ampere, one should not be altogether surprised that power loss, even within large neurons, is minimal.[4]

This brings us neatly on to another fundamental consideration of conductors - or resistors as you prefer - one which affects both the transmission of electrical current from a power station to a consumer and the transmission of a nerve impulse along the axon of a spinal motor neuron. A long conductive cylinder, be it a metallic cable or an axon, which is surrounded by an insulating sheath of significant electrical resistance, such as a myelin coat or a plastic cover effectively forms a coaxial cable. A nerve axon comprises a conductive cytoplasmic gel which is enveloped by a long insulating myelinated membrane sheath (contributed by specialised glial cells known as Schwann cells) which affords a far higher electrical resistance than the gel it encompasses. Thus, we can estimate how quickly a given voltage V_o will decay across the length of the axon if we can determine its length constant (λ).

---

[4] Where energy loss is likely to occur is in the dissipation of those ionic gradients that support electrical activity within excitable tissues, as the diffusion of ions down their respective concentration gradients requires the expenditure of considerable chemical energy to reverse.

The length constant (λ) effectively serves as an electrical measurement of the distance that a given voltage (V_o) decays to $1/e$ - fold of its original value.

Based upon cable theory, we can calculate λ by determining the ratio of the membrane resistance (R_m) to the axial resistance of the cytoplasm of the axon (R_a), where;

$$\lambda = \sqrt{(Rm/Ra)}$$

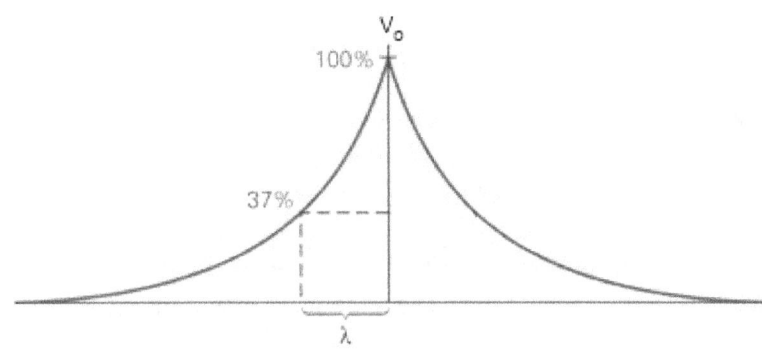

Thus, broader nerve processes with a greater axial diameter have a longer length constant (*i.e.* a more gradual decay in voltage) than smaller, finer axons, as the higher the ratio of Rm to R_a, the less quickly a given voltage attenuates along the length of the axon. Thus, myelination and the creation of miniature electrical substations along the length of the axon, known as *nodes of Ranvier*, serve to maintain the integrity of the signal and the distance over which nerve impulses may be faithfully maintained. However, this is only a cursory consideration of cable theory, and we will not delve further into the complexities of myelinated axons within this book.

It might be argued, unless we are discussing an electric eel which is capable of discharging currents at 860V, that such theoretical considerations as power loss and 'electrical fields' may seem somewhat redundant in relation to biological systems, but we should remember that the fundamental principles of electricity apply, even at the cellular level.

A brief introduction to membrane capacitance

The cell membrane is, in essence, a near perfect capacitor. Its ability to store electrical charge enables it to drive a variety of excitable processes on command via the opening of ion channels which are embedded within it.

So why is the cell membrane a perfect capacitor? To understand this, we need to return to the basic determinants of an ideal capacitor. In its purest form, a capacitor is created when two conductive surfaces are separated by an insulating or non-conductive medium. Capacitance (C) is given by;

$$C = \varepsilon r . \varepsilon 0 \left(\frac{A}{d}\right)$$

where A is the overlapping area of the two membrane plates; (d) is the distance that separates them; and ε_r is the permittivity (or willingness to conduct electrical charge or current, if you prefer) of the intervening medium. A perfect vacuum is a fairly effective insulator and is given an ε_r of 1. ε_0 is the all-important *dielectric constant* (which has a value of 8.854×10^{-12} F.m^{-1}).

As we may deduce from the preceding equation, the thickness of the cell membrane, which is typically less than 10nm, makes it an ideal capacitor, as does the permittivity of the lipid bilayer [5] which forms it which, with an estimated value of 2 [6] approaches the value for a perfect vacuum.

While the solid-state electronics industry produces bulky capacitors that can store many Farads of charge, the fine bilayer of the cell membrane has a *specific* capacitance which has been estimated at 1μF/cm^2 [7], thereby providing a store of electromotive force which is available on demand.

So why is a preliminary discussion of membrane capacitance a prerequisite for the chapters that will follow? Other than the fact that it enables cells to store electrical charge, we need to understand the fundamental electrical properties of cells when we attempt to record from them using glass electrodes.

---

[5] Note that the lipid bilayer comprises a polarised arrangement of hydrophilic heads, which may carry a charge, and hydrophobic tails.
[6] W Huang and D G Levitt, Theoretical calculation of the dielectric constant of a bilayer membrane. Biophysical Journal. 1977 Feb; 17(2): 111–128.
[7] Luc J. Gentet, Greg J. Stuart, & John D. Clements. Direct Measurement of Specific Membrane Capacitance in Neurons. Biophysical Journal. Volume 79, Issue 1, July 2000, Pages 314-320.

Recording from cell membranes with electrodes

One curious, yet highly beneficial property of polished glass is that it has an *extremely high affinity* for the cell membrane (if both are clean). In fact, the resulting electrical resistance of this bond between a smooth, polished glass surface and a cell membrane may exceed one million, million Ohms (>100GΩ). This effectively enables us to electrically isolate the ion channels within the cell membrane and to record the tiny currents that flow across them.

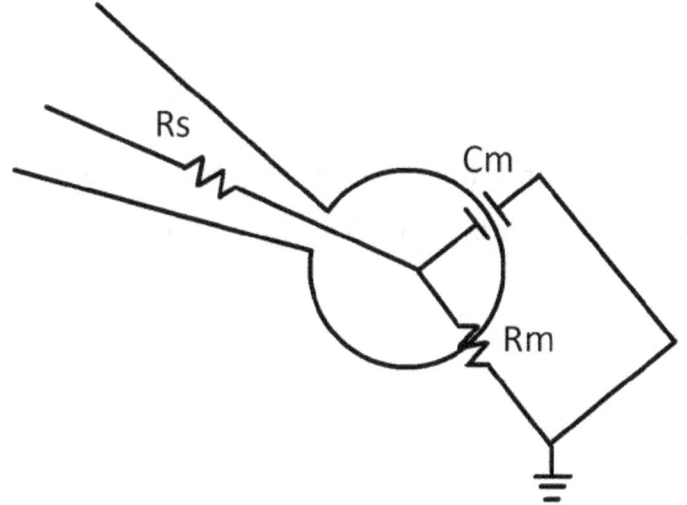

In the equivalent electrical circuit presented *(inset left)* we can see that when we record from a cell membrane we are effectively dealing with an RC circuit or, in other words, a resistor which is in parallel with a capacitor. In this instance, the membrane capacitance (C_m) is represented by two parallel plates (| |), while the membrane resistance (R_m) is represented by the zig-zag line.

However, there are several important considerations which deviate from our conventional understanding of solid-state RC circuits.

[1] Ion channels are constantly opening and closing, which means that R_m is not a constant.

[2] If the conductance of the cell membrane is large (*i.e.* the cell has an input resistance of less than 100 MΩ), then the *series resistance error* arising from the resistance between the electrode and the cell interior (R_s) becomes a significant issue, as the command voltage (V_c) imposed by the electrode will be divided across the two resistances R_s and R_m which are effectively in series. Thus, $V_c = V_s + V_m$, where the voltage-drop across each resistance R_s and R_m is proportional to the fraction of their combined resistance (R_t).

For instance, if we impose a hypothetical command voltage of 100 mV across a cell membrane using an electrode with a series resistance of 10 MΩ and the cell has an input resistance (R_m)

of 90 MΩ, then a voltage drop (Vm) of only 90 mV will be imposed across the cell membrane (*i.e.* there is a series resistance error of 10%). This is an extreme example, as most cells have larger input resistances, although the series resistance error should always remain a consideration.

[3] Cells do not usually have their membrane potential 'voltage clamped', but rather the membrane potential changes dynamically in response to the flow of ionic currents across it. However, the voltage clamp remains a useful tool in that it allows us to study the cell as an equivalent electrical circuit and to observe the behaviour of ion channels at a range of imposed potentials, creating an electrical signature or 'fingerprint.'

[4] If we are recording from more than one ion channel, which is usually the case, the total current flowing through the membrane (*Im*) is given by the following key equation.

$$Im = i.n.Po$$

where, for any given membrane potential, i is the unitary current flowing through each open ion channel; n is the number of channels active within the cell membrane, and Po is the statistical probability of finding the channel open at a given membrane potential, or '*open probability*'.

When we study currents across the whole cell membrane, we are inevitably curious about how each of these parameters shapes the currents flowing across the whole cell membrane. As individual ion channels are constantly opening and closing, studying their aggregate behaviour guides our understanding of their physiological function.

[5] When we charge *(or discharge)* an RC circuit by imposing *(or removing)* a voltage step, then a characteristic exponential time course is observed as the capacitor charges *(or discharges)*. This also holds true when we apply a command potential across a cell membrane or inject current into a cell. The time course of this charging process is described by the following equation;

$$\tau = Rm.Cm$$

where, τ is the time constant required for the charge (measured as an electrical potential difference) on the capacitor to reach 63% of its peak value. However, as we shall soon discover, in the world of biology (unlike solid-state electronics), such parameters as *Po, Rm,* and *Cm,* are anything but constant…

As we progress through the practical problems presented within this book, we shall introduce further conceptual challenges to our understanding of the complex world of ion channel kinetics.

However, such journeys are always best undertaken one bridge at a time…

Chapter Two

Probing an electrical black box

For our first exercise, we shall begin by examining the output of the whole cell rather than studying the properties of the constituent ion channels within the membrane. As a fledgling Ph.D. student at Cambridge University, I was presented with an unknown - an electrical black box which had never been recorded from before - the small intestinal crypt.

The question that remained unanswered was whether the small intestinal crypt, an almost perfect cylinder composed of epithelial cells, possesses the necessary arrangement of ion channels to effect the secretion of fluid and electrolytes in response to neurotransmitters.

To address this question, I had to probe this enigmatic electrical black box using some original applications of classical biophysical principles.

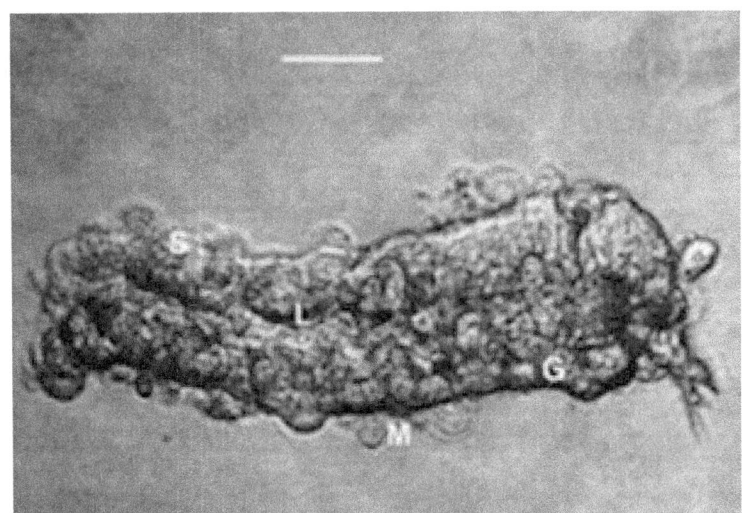

The biophysical question

The epithelial cell monolayer lining the cavity, or *lumen* of the small intestine mediates both the absorption and the secretion of fluid and electrolytes. Although the net **absorption** of fluid and electrolytes normally prevails, facilitating the uptake of simple sugars and amino acids are coupled to transepithelial Na^+ movements, the small intestine is also capable of the net secretion of fluid and electrolytes.

An isolated small intestinal crypt may be viewed above (the scale bar represents 100μm). Note the important anatomical features. The crypt is a cylinder which is formed from a single layer of epithelial cells, or *enterocytes*, which is enclosed at one end and opens into the lumen (L) at the other. This creates two anatomically distinct surfaces. The luminal surface, which is inaccessible in the intact crypt, opens into the cavity of the small intestine, while the basolateral surface (S) is accessible to blood vessels, nerve terminals, *and* electrodes. The middle (M) and enclosed ground (G) regions of the crypt are also labelled.

The vectorial transport of ions across polarised epithelial cells

To achieve the net transport of fluid and electrolytes across the intestinal epithelium, a selective permeability barrier must be generated which regulates the exchange of ions and nutrients between the lumen and the body. By functionally dividing the plasma membrane into lumenal and basolateral surfaces and by asymmetrically distributing membrane pumps and ion channels, epithelia are capable of mediating the net transport of fluid and electrolytes.

The small intestinal epithelium is referred to as a *'leaky'*, or low resistance epithelium, because it possesses an intercellular pathway which remains highly permeable to water and cations, especially Na^+. This paracellular pathway arises because the individual crypt enterocytes are joined by proteins known as tight junctions.

It had long been established that the absorption of nutrients is mediated by the villus enterocytes which project out into the lumen of the small intestine. However, if the small intestinal crypt *is* capable of mediating the secretion of fluid and electrolytes, then it must possess all the elements of a biological battery, with a polarised distribution of ion channels that are selective to Cl- within its luminal, or apical membrane, and a concentration of cation channels within its basolateral membrane which would provide the electromotive force to drive Cl- efflux. In theory, if ion channels that were selective to Cl- were to open and Cl- ions left the cell across the luminal membrane then Na+ ions would follow via the paracellular pathway, driven by a transepithelial potential difference. Where the salt flows, so the law of osmosis dictates, water is obligated to follow…

This is summarised in the equivalent circuit diagram which may be found inset, where the apical (a), or luminal, and basolateral (b) membranes are each modelled as a distinct RC circuit with an ionic conductance in parallel with a membrane capacitance. Note that the *variable* apical Cl- conductance (g_{Cl-}) is theorised to be activated by neurotransmitters and therefore rate-limiting. This is denoted by a switch. The basolateral membrane is represented by a variable cationic conductance (g_{cat+}) and the paracellular pathway as a parallel conductance (g_{Na+}).

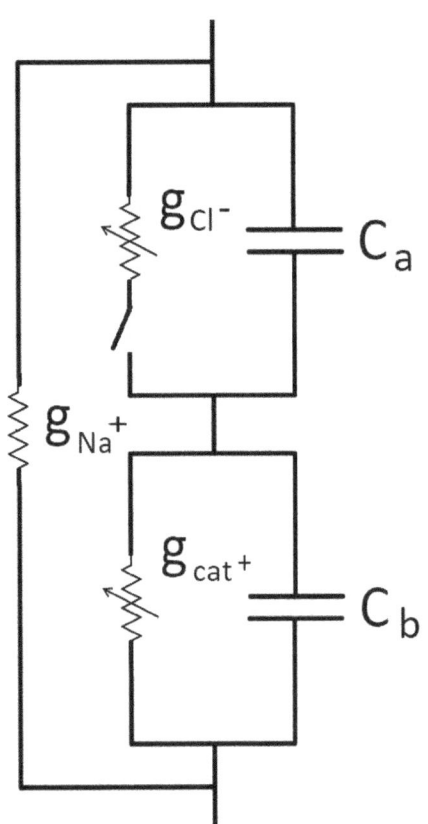

This asymmetrical arrangement of ion channels would enable the crypt epithelium to effect the unidirectional transport of electrolytes and water in response to neurotransmitters released by the intestinal nervous system, or *myenteric plexus*. Neurotransmitters which are known to evoke small intestinal secretion include acetylcholine

(ACh) and Vasoactive Intestinal Polypeptide (VIP). In fact, both of these key neurotransmitters, which are present throughout the central nervous system, were originally discovered in the small intestine.

Investigating the electrical properties of the crypt using the patch-clamp technique

The discovery that polished glass electrodes form high resistance electrical seals when brought into contact with the cell membrane earned Erwin Neher & Bert Sakmann the Nobel Prize.[8] Their pioneering development of the patch clamp technique has made it possible for us to record from individual ion channels by amplifying the tiny currents which flow through them.

The astonishing electrical resistance of the seal which forms between the pipette tip and the membrane surface, a value which can exceed 100GΩ, confers fundamental advantages. The GΩ seal ensures that currents traversing the membrane patch flow almost exclusively into the patch pipette, thereby minimising the arch enemy of all electrophysiologists, namely unwanted signal or 'bad' noise. Bad noise derives from unwanted electromagnetic signals within the background as well as from that thermal noise which arises due to the random motion of excited charge carriers within a conductor. While it is important not to filter out the 'good' noise that arises from the opening and closing of ion channels (*i.e.* the signal we are trying to record), leaky seals are especially noisy and effectively preclude the resolution of individual ion channels.

While simply placing a polished electrode against the membrane surface of a cell allows us to record the currents flowing across that isolated 'patch' of membrane, if we rupture this patch by suction or with pore-forming antibiotics then we can record either the currents flowing across the entire cell membrane by clamping the membrane voltage, or the changes in membrane potential that arise by 'clamping' the current flowing into the patch pipette.

---

[8] Erwin Neher & Bert Sakmann, 'Single-channel currents recorded from membrane of denervated frog muscle fibres.' Nature 260, 799 - 802 (29th April 1976).

We shall begin with a series of classical biophysical experiments which are designed to determine the ionic permeability of the basolateral membrane.

Effects of ion substitution experiments upon crypt membrane potential

After GΩ seals were obtained with nystatin-filled electrodes,[9] the membrane potential (Em) was monitored in the *current-clamp* mode of the patch-clamp amplifier *(see Appendix A for experimental solutions and methods).*

To test the dependence of Em upon extracellular ion concentration, a series of ion substitution experiments were performed in which the ratio of cations or anions were changed without changing the osmolarity of

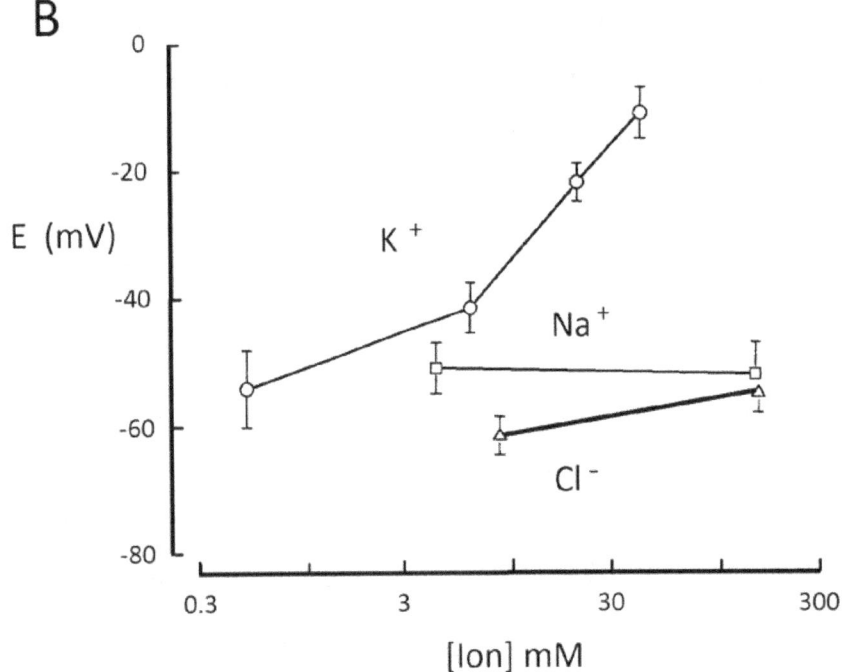

---

[9] Nystatin is a polyene antibiotic which forms pores within cell membranes which are permeable only to monovalent ions. This is the so-called "perforated" patch clamp technique which enables us to record electrical currents without the loss of intracellular contents.

the bath solution.[10]

The figure inset *(above left)* shows a recording obtained from a small intestinal crypt which was originally bathed in a normal 'physiological' recording solution with an extracellular K+ concentration [K+]$_o$ of 6.2mM and an initial Em of -64mV.

When we increase the extracellular K+ concentration first to 20 and then to 40mM (by the equimolar substitution of NaCl with KCl), we elicit a rapid depolarisation of the membrane potential which is readily reversible (A). Conversely, decreasing the extracellular K+ to 0.5 mM elicited a hyperpolarization of crypt Em.

Please note that, by convention, membrane depolarisations are presented as upward deflections and hyperpolarisations as download deflections of the membrane potential recording.

However, when all but 8mM of the Cl- present within the bathing medium was replaced with the bulky, impermeable anion gluconate⁻, only a small hyperpolarization was evoked; whereas replacing extracellular Na+ with the impermeant cation N-methyl-D-glucamine (NMDG$^+$) was virtually without effect.[11]

When we present (B) the average changes in membrane potential (E_m) elicited by these ionic substitutions as a function of [ion]$_o$ concentration, it is evident that the resting membrane potential of the crypt is strongly dependent upon the extracellular K+ concentration, with a slope equivalent to a 37mV change in E_m for every ten-fold change in the extracellular K+ concentration. This approaches the 58mV change that would be hypothetically predicted by the *Nernst equation* for a membrane conductance that was exclusively permeable to K+ ions. When K+ channels open they tend to drive E_m towards the equilibrium potential for K+, which we term E_K. Thus, E_K serves as an effective 'lower limit' for E_m and any increase in the K+ conductance will tend to hyperpolarise the membrane potential of the small intestinal crypt.

---

[10] This would cause changes in cell volume and activate ion channels whose function it is to maintain cell volume and intracellular osmolarity.

[11] Note that the inaccessibility of the apical membrane due to the lumen (and any tonic level of secretion that may occur) precludes the determination of which, if any, apical conductance(s) may contribute to the spontaneous resting membrane potential.

The Nernst equation

This brings us to one of the central theoretical principles of electrophysiology which was first described by Walther Nernst. The Nernst equation enables us to predict the precise potential difference that will arise across a cell membrane based upon the charge (*i.e.* valence) and the concentration gradient for a given ionic species across it (if the membrane is perfectly selective for that ion). The so-called Nernst potential for any given ion is that membrane potential at which the ion is said to be in equilibrium, *i.e.* there is no net movement of the ion across the membrane.

Thus, the Nernst potential (V) for any given ionic species is more commonly referred to as the *equilibrium potential* for that ion. The equation for a single ionic species, in this case K+, is presented in the following form;

$$V = \frac{RT}{zF} \ln \frac{[K+]o}{[K+]i}$$

where R is the universal gas constant (8.314 Joules per Kelvin per mole); T is the bath temperature in Kelvin; z is the charge or valence of the ion (+1 for K+), and F is the Faraday's constant (or 96,485 Coulombs per mole). $[K+]_o$ denotes the external K+ concentration and $[K+]_i$ the intracellular K+ concentration, which is an unknown. However, if we can measure the membrane potential of the cell and effectively control the extracellular K+ concentration, then we can estimate the intracellular K+ concentration using the Nernst equation.[12] When more than one type of ionic species is involved, as is usually the case, the relationship becomes more complex and brings us to the Goldman-Hodgkin-Katz equation, which we will address within the next section.

---

[12] The concentration of K+ ions inside the cell is invariably far higher than it is outside, courtesy of a very important pump known as the Na+-K+ ATPase which exports 3 Na+ ions from the cell interior in exchange for every 2 K+ ions that it imports to create a membrane potential which is negative in all resting 'excitable' cells. Thus, the activation of Na+ selective channels will trigger an influx of Na+ ions into the cell and tend to depolarise the cell membrane and an activation of K+ channels will trigger an efflux of K+ ions and thus evoke a hyperpolarisation. The 'electrogenic' Na+-K+ ATPase then seeks to restore the balance.

The Goldman-Hodgkin-Katz equation

Under most true physiological conditions, the resting membrane potential of an 'excitable' cell will be determined by the equilibrium potential of more than one ion and will lie between the equilibrium potentials for each ion to which the membrane is permeable. As you might reasonably surmise, based upon the preceding experiment, the more permeable a membrane is to a particular ion, the closer to its equilibrium potential the membrane potential will be.

This is described by the Goldman-Hodgkin-Katz equation which is usually presented in the following form (for two permeant ions);

$$Vm = \frac{RT}{zF} \cdot \ln\left(\frac{P_{Na}.[Na+]o + P_K.[K+]o}{P_{Na}.[Na+]i + P_K.[K+]i}\right)$$

where V_m is the membrane potential observed and RT/F may be considered a constant for a given bath temperature. The term P denotes the permeability of the membrane to a given ion.

Despite its outwardly fearsome appearance, the Goldman-Hodgkin-Katz, or GHK equation is very useful in predicting the outcome of any changes in ion channel activity. For instance, if there is no permeability to Na+ (or to Ca++), then these terms simply disappear from the equation, simplifying matters considerably.

The GHK equation simply serves to state the fact that the relative weight, or permeability to a given ion, determines the membrane potential of the cell. Thus, the resting membrane potential of the small intestinal crypt, which is dominated by a basolateral K+ conductance, approaches the expected equilibrium potential for K+, or E_K.

If on the other hand, we excite a cell by opening Na+ channels, then the membrane potential will 'depolarise' rapidly towards the equilibrium potential for Na+. As we will later come to realise, Cl- ions are the great variable of cellular electrophysiology.

Whole cell currents recorded from unstimulated small intestinal crypts

Having begun proceedings in the current clamp mode of the patch clamp amplifier, we shall now revert to the more popular, although less 'physiological' voltage clamp mode. We shall seek to record the currents which are present within the resting membrane of the whole crypt across a range of imposed voltages. These experiments provide us with our first unexpected surprise...

For larger, intact small intestinal crypts we are simply unable to clamp or hold the membrane potential of the crypt. This immediately informs us that, far from acting alone, the individual crypt enterocytes are *acting in concert*. This infers that the individual crypt enterocytes are electrically coupled by a family of ion channels known as *gap junctions*. We will address their importance again later when we record from the insulin-secreting β-cells of the pancreas. For now, we will simply state that gap junctional channels serve as conduits between adjacent cells and that this means that the cells of the crypt, like those of the heart, act in concert as an electrical *syncytium*.

However, it did prove to be possible to record the whole cell currents from some smaller crypts that had been fragmented by the isolation procedure. A series of incremental holding potentials was imposed upon three crypt fragments from a 'holding potential' of -40mV. Whole cell current recordings using nystatin-perforated patches could only be obtained from crypts when the measured capacitance compensation did not exceed 100pF. Here we address one of the limitations of the patch clamp technique – the practical ability to spatially 'clamp' the voltage across the cell membrane - a problem that is routinely encountered when we attempt to record from very large cells, those which are electrically coupled, or neurons which contain long, fine processes. This issue is addressed in more detail within Appendix C.

A series of voltage steps of 200ms duration were applied in 40mV increments from -140 to 100mV as shown in panel B below. This voltage protocol elicited the currents shown in panel A. The resulting relationships between the average current and holding potential for three crypts are shown in panel C.

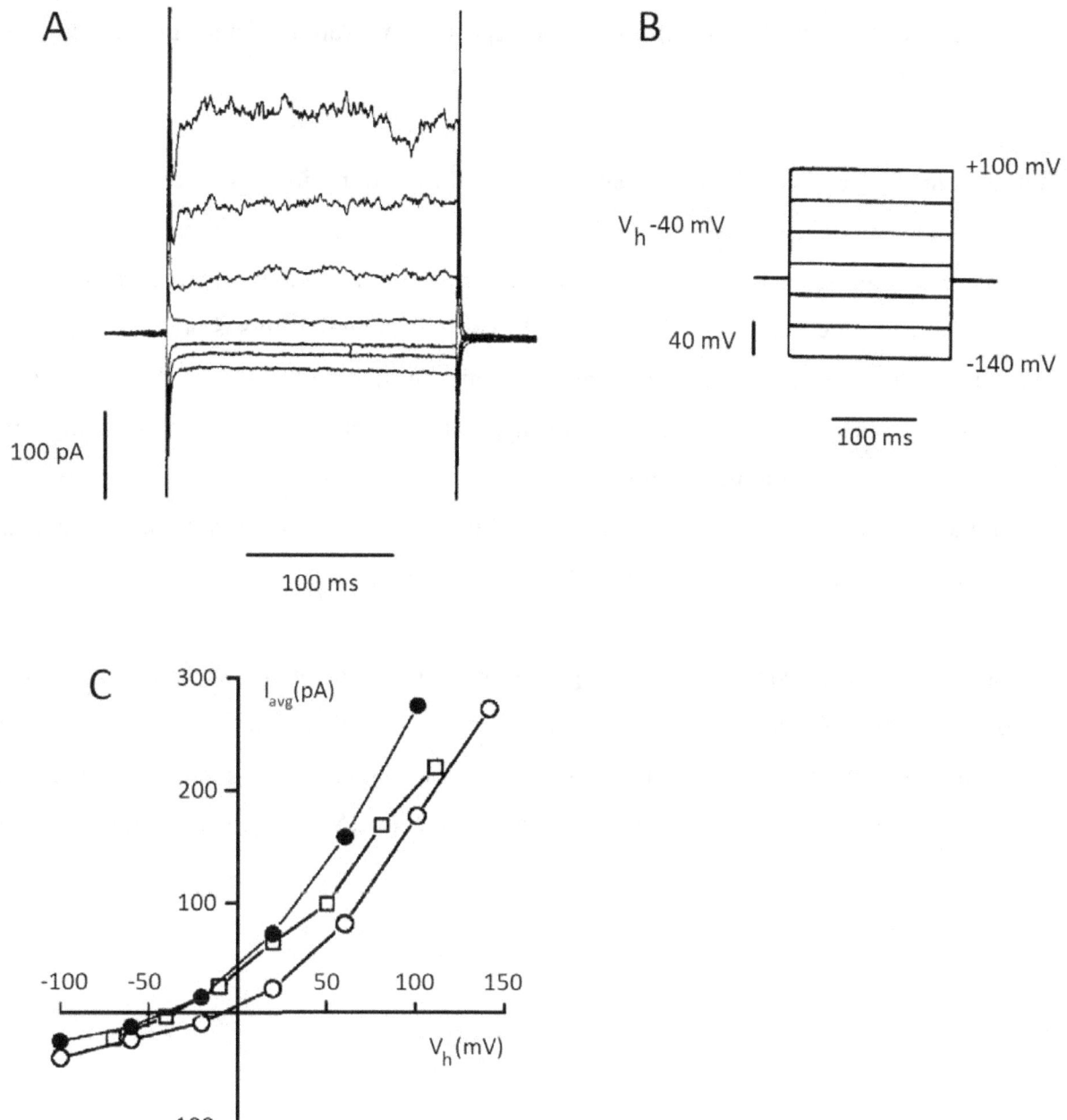

Note that, by convention, outward currents are presented as upward deflections, and inward currents as download deflections. The rapid spikes elicited on the onset and offset of voltage pulse applications are caused by the charging and discharging of the pipette capacitance.

We can clearly see that the currents recorded appear to show a marked outward rectification (panels A & C) as well as an apparent time-dependent activation of outward currents at more depolarised holding potentials.

Although the gradient of the current-voltage (I-V) relationship gets steeper as the holding potential becomes more positive, in other words, they are said to be 'outwardly rectifying', this may be explained, at least in part, by the greater currents that would be predicted by the asymmetrical concentrations of ions across the cell membrane. If this is predicted by the GHK equation, then it is termed 'Goldman rectification', but if it deviates from the predicted value because of changes in single channel open probability or the relative permeability of the membrane to different ions, then it may be regarded as being non-Goldman rectification. Given that there is clearly an activation of the outward current at more depolarised holding potentials, then the rectification cannot be fitted to the GHK equation.

Thus, we can conclude that the resting membrane potential of the small intestinal crypt is dominated by a K+ conductance which is present within the basolateral membrane which is outwardly-rectifying and activated by membrane depolarisation. This will later serve as a 'kinetic fingerprint' when we investigate the properties of the constituent ion channels of the basolateral membrane.

Chapter Three

Exciting the membrane

It has long been established that a number of neurotransmitters which are released from the nerve terminals that innervate the small intestinal crypt stimulate the secretion of fluid and electrolytes within the small intestine. We selected two of these to study, specifically acetylcholine and VIP, as they are known to bind to receptors which govern very distinct intracellular signalling pathways.

When VIP binds to its receptor on the crypt basolateral membrane it activates an enzyme known as adenylate cyclase which increases intracellular cAMP levels, whereas acetylcholine (and its analogue carbachol) bind to another class of receptor known as *muscarinic* receptors which cause calcium to be released from intracellular stores.

Thus, I applied each of these two receptor agonists, in turn, to determine what would happen to the ionic conductances of the small intestinal crypt when they were excited via these two very different signalling pathways, in essence seeking to mimic the action of the endogenous neurotransmitters. Again, the more physiological current clamp mode of the perforated patch clamp technique was employed to study any resulting changes in membrane conductance.

The effect of carbachol upon crypt membrane conductance

The effect of *muscarinic* receptor activation on crypt E_m was determined, and the recording shown below (panel A) is of a crypt with an initial E_m value of -57mV. The addition of 100mM carbachol evoked a 15mV *hyperpolarization* (*i.e.* towards the equilibrium potential of K+, or E_K), which was sustained for the duration of agonist application, and was readily reversed upon washout.

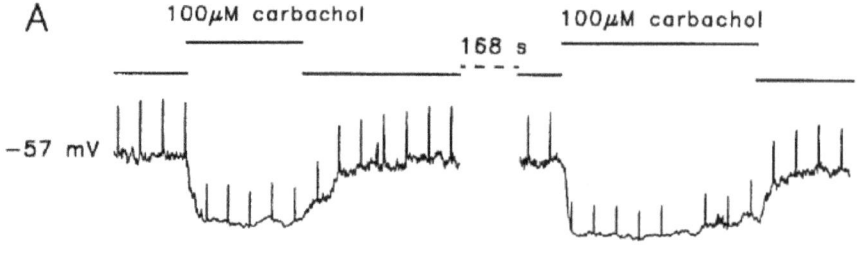

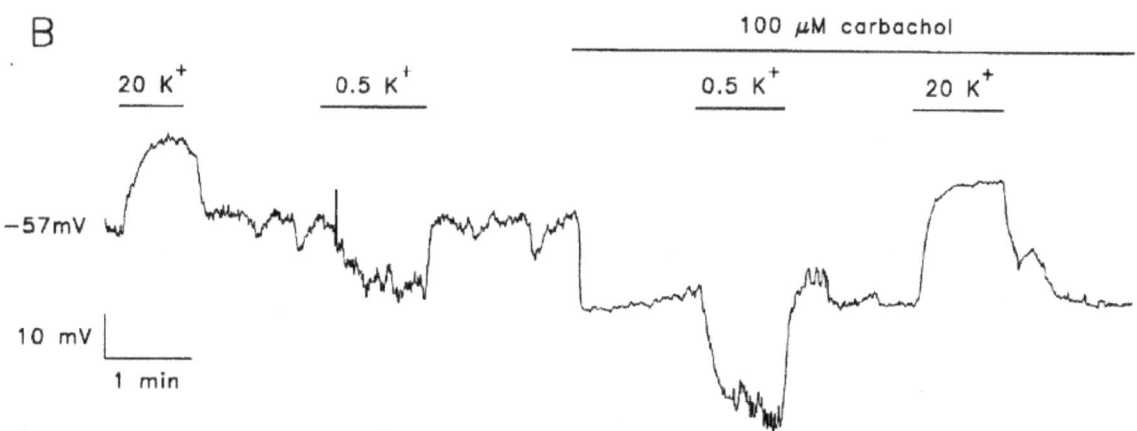

Note that I was applying a train of small current pulses throughout the recording to probe for underlying changes in membrane conductance as, for a given current pulse, the input resistance of the membrane will be given by a simple rearrangement of Ohm's Law, where;

$$Rm = \frac{\Delta V}{\Delta I}$$

Thus, as the size of the deflection in E_m in response to a train of constant current pulses *decreases* during carbachol application, this indicates that the hyperpolarization of the crypt membrane potential *towards* E_K is associated with a small *increase* in membrane K+ conductance. As carbachol is known to increase intracellular levels of Ca2+ in many excitable cells, this raises the intriguing possibility that we are observing the activation of a novel K+ channel. The Ca2+ dependence of this carbachol-activated K+ conductance was confirmed by applying carbachol in the presence and absence of extracellular Ca2+ (*not shown here*).

To further test this hypothesis, I investigated the ionic basis underlying the carbachol-induced hyperpolarization, by changing the extracellular concentration of K+ in the presence and absence of the agonist. Panel B of the figure above shows the effect of changing $[K+]_o$ first to 0.5 and then 20mM under control conditions, which produced the hyperpolarization and depolarisation of E_m as observed in the previous chapter. However, the subsequent addition of carbachol produced a sustained hyperpolarization during which the magnitude of the deflections in E_m in response to changes in extracellular $[K+]_o$ were clearly increased, indicating that *p*K had increased.

Note that, by applying a series of current pulses to the membrane we are effectively briefly charging (*and discharging*) the membrane capacitance. We can, in theory, use these to derive the time constant τ by calculating the input resistance (R_m) from the evoked changes in E_m by a series of current pulses of a known amplitude (I), given that;

$$Rm = \frac{\Delta Em}{\Delta I}$$

From a determination of R_m, we can estimate crypt membrane capacitance (C_m) from;

$$\tau = Rm.Cm$$

The effect of VIP upon crypt membrane conductance

Next, I sought to determine the effects of the cAMP mobilizing agonist VIP upon crypt membrane conductance. In the following figure, we observe a continuous recording of crypt E_m during which 50nM VIP was added to the bathing medium for the duration indicated by the solid bar. After a short delay, VIP evoked a 30mV *depolarisation*, which was *associated with a substantial increase in membrane conductance*. This could only rationally be caused via the activation of either a membrane Na+ or Ca2+ conductance, or by the activation of a Cl- conductance (the efflux of anions is electrically equivalent to the influx of Na+ ions) **if Cl- is maintained above electrochemical equilibrium** within the small intestinal crypt.

Further, the activation of Na+ or Ca2+ channels would be expected to cause an increase in the volume of the cell, as the free concentration of these ions is considerably greater outside the cell than inside, an electrochemical gradient which is actively maintained by the all-important Na+-K+ ATPase.

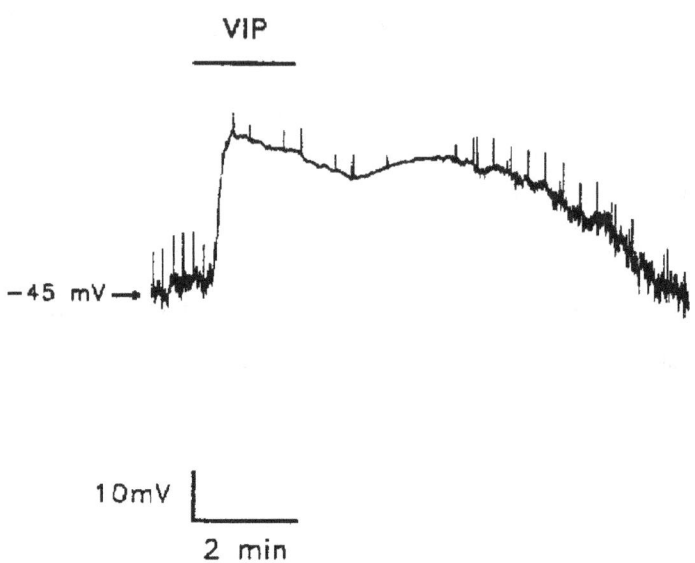

Another method by which we can test the ionic nature of the membrane conductance underlying these depolarisations is to c*hange the concentration of Cl- within the patch pipette* as we are, in effect, equilibrating the intracellular concentration of Cl- with the comparatively far greater volume of the salt solution (KCl) contained within the electrode. By applying VIP with two substantially different concentrations of Cl- in the electrode, if the depolarisation of crypt membrane potential arises through the activation of a membrane Cl- conductance then, if we change the equilibrium potential of Cl- (E_{Cl}), we should theoretically change the magnitude of the depolarisations evoked.

In four separate experiments, the depolarisations evoked by VIP were significantly smaller with 60mM Cl- in the patch pipette (24 ± 2mV) than with 145mM Cl- (32 ± 1mV), consistent with the predicted E_{Cl} values of -23mV and 0.4mV, respectively *(assuming the complete equilibration of the intracellular milieu with the pipette solution)*.

The next question was whether the effect of VIP might be mediated via the activation of the enzyme adenylate cyclase and an increase in levels of the second messenger cAMP.[13] To address this question we applied the phytochemical forskolin, a naturally occurring activator of adenylate cyclase. As we can see from the recording below, the addition of forskolin [FSK] mimics the depolarising effect of VIP *(although, unlike VIP, its activation of adenylate cyclase is irreversible and cannot simply be 'washed' from the system)*.

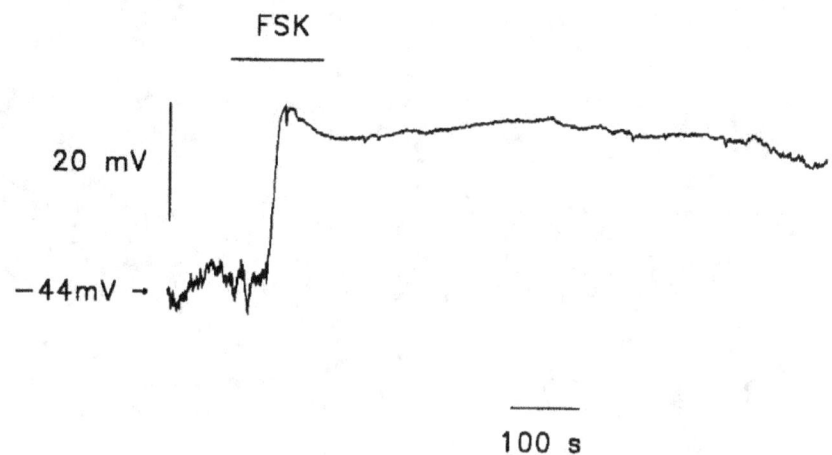

But what of our previous contention that the influx of Na+ and/or Ca2+ channels would be expected to cause the volume of the crypt to swell? Surely then, the converse would also hold true, and the activation of Cl- channels would evoke an efflux of Cl- from the crypt and cause a decrease in crypt volume? This hypothesis was tested when we applied VIP to a crypt and captured a series of images immediately before (*see* panel A below) and 8 minutes after (B) the addition of 100nM VIP to the bath.[14] Clearly, a significant decrease in crypt volume is apparent, and this can be estimated by assuming that the volume of a crypt (V) can be approximated to that of a cylinder, which is given by the equation;

$$V = \pi.r^2.h.$$

---

[13] An activator of Protein Kinase A and the Cl- channel known as the Cystic Fibrosis Transmembrane Conductance Regulator (CFTR) which was subsequently shown to be present in small intestinal crypts (Trezise AE, Buchwald, M. In vivo cell-specific expression of the cystic fibrosis transmembrane conductance regulator. Nature. 1991 Oct 3;353(6343):434-7.) *These experiments were conducted between April and September 1991.*

[14] The scale bar represents 50μm and the 31% decrease in volume can be estimated by calculating the volume of a cylinder = $\pi.r^2.h$.

Both VIP and carbachol evoked this volume decrease, which could be prevented by raising the extracellular K+ concentration (*i.e.* eliminating the electrochemical driving force for K+ efflux from the crypt) and also by known inhibitors of Cl- channels. Thus, we can deduce that both VIP and carbachol stimulate the loss of KCl (and associated water) from the crypt via the activation of either a K+ conductance (carbachol) or a Cl- conductance (VIP).

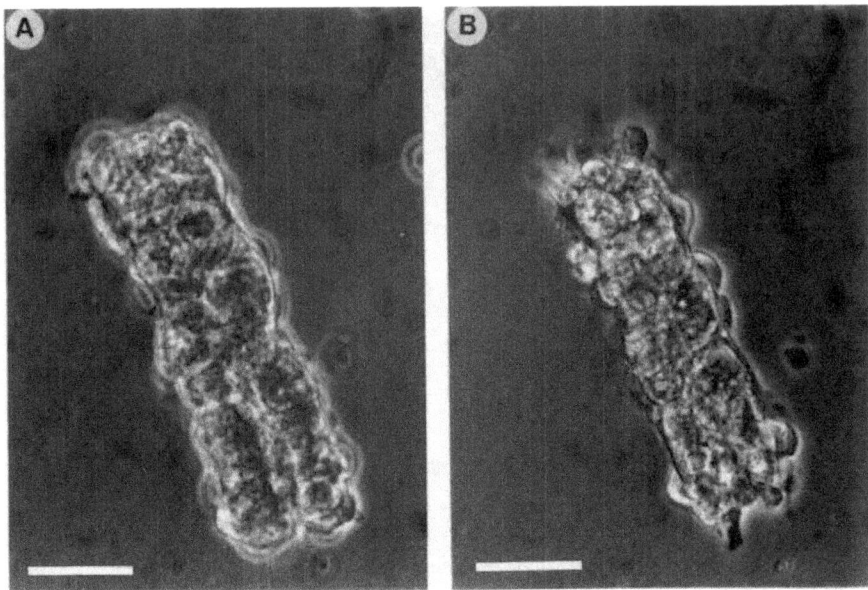

Could this mean that the actions of VIP and carbachol might be cooperative (i.e. synergistic) in nature, as they appear to have very distinct actions upon the membrane conductance of the small intestinal crypt?

This possibility is highlighted within the figure below, in which I applied a 'ramp' potential within the voltage clamp mode to derive a *direct I-V relationship* from both the resting crypt and in response to the application of carbachol or forskolin. The effect of the application of forskolin upon the whole cell current elicited by a linear ramp potential (B) is shown below.

Note that forskolin increased the membrane conductance of the crypt *and* shifted the reversal potential from -39mV to -11mV, which was associated with a large increase in both the inward and outward currents at negative and positive holding potentials, respectively; whereas the application of carbachol shifted the reversal potential towards E_K, as predicted, and increased the size of the outward K+ current only.

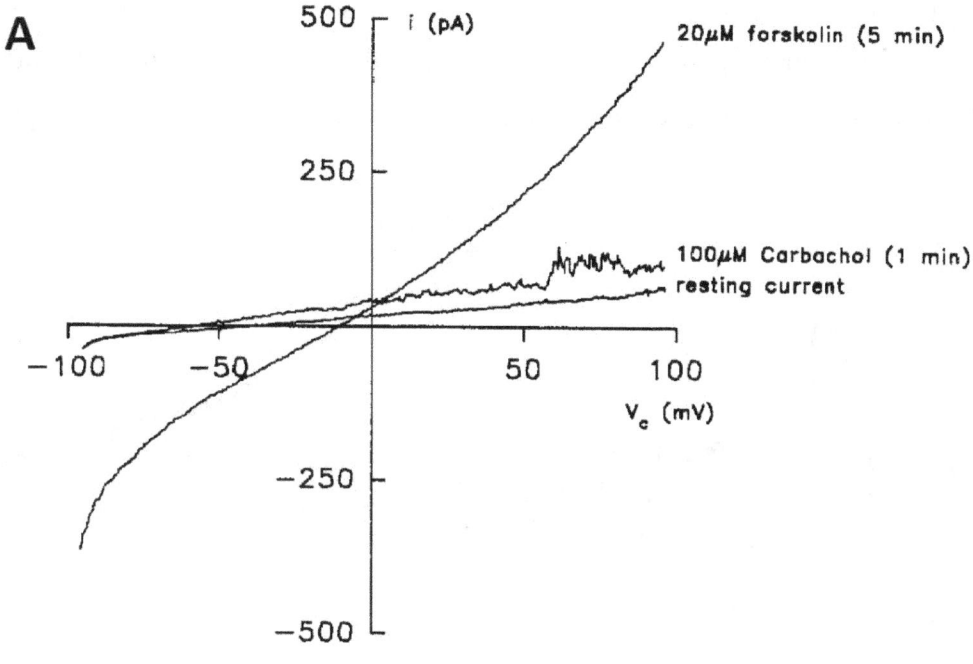

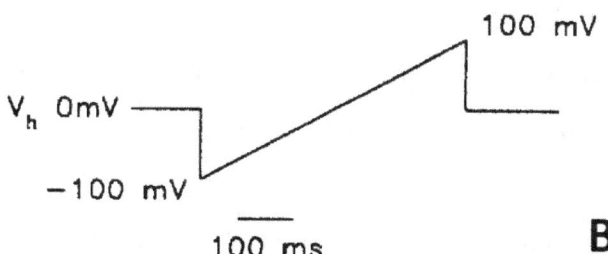

Although applying such a brief ramp potential is *entirely unphysiological*, it is nonetheless a useful technique in that it affords a 'snapshot' of the reversal potential of the cell membrane whose value is predicted by the GHK equation.

For our final insight, we asked whether VIP and carbachol might act synergistically in exciting the membrane conductance of the small intestinal crypt?

The effects of 100nM VIP and 100mM carbachol (CCH) upon crypt membrane potential (E_m) are shown in the figure *inset*, when applied individually and in concert. In the presence of VIP alone, a characteristic depolarisation was observed, which recovered to the resting value of Em within 6 minutes of washout. The subsequent addition of carbachol produced a characteristic hyperpolarization which was again rapidly reversed upon washout. However, the simultaneous addition of carbachol *and* VIP, after inducing a transient hyperpolarization,

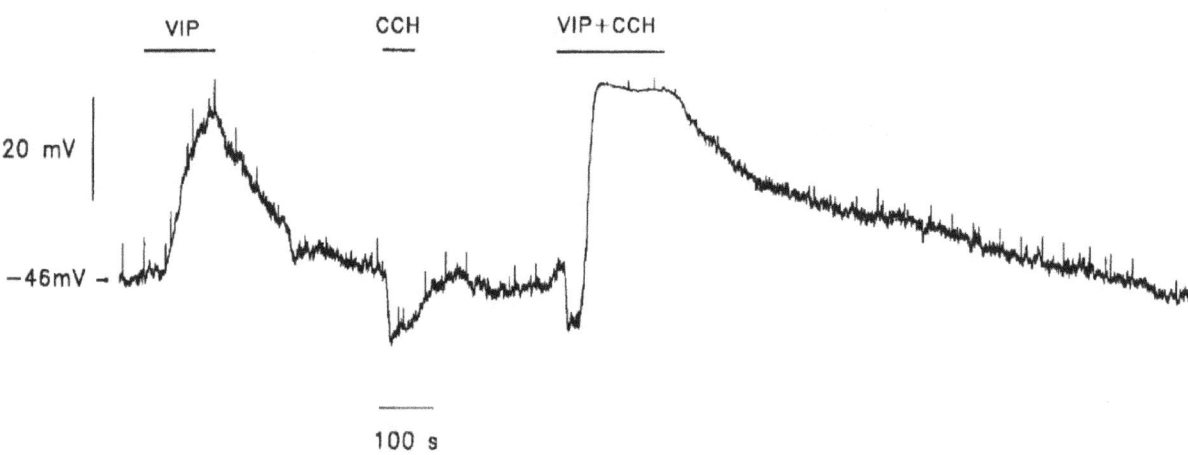

then evoked a marked depolarisation which was larger than that produced by VIP alone. Further, the deflections in E_m in response to a constant train of current pulses indicate that the increase in membrane conductance evoked by both agonists was greater than for the addition of either agonist alone.

Therefore, we can conclude that muscarinic and VIP receptor stimulation both evoke conductance changes in isolated small intestinal crypts which are consistent with the activation of K+ and Cl- conductance pathways, respectively.

The experimental data is consistent with the hypothesis that the small intestinal crypt mediates fluid and electrolyte secretion. Quod erat demonstrandum.

In the following chapter, we shall delve more deeply into the question of chloride equilibrium potentials - the great unknown of cellular electrophysiology.

Chapter Four

The enigma of E_{Cl}

In this chapter, we shall explore the importance of Cl- equilibrium potentials in electrophysiology. In particular, we shall address the role they play in determining the physiological function of excitable cells, an aspect that was originally overlooked by classical investigators of the polarized world of action potentials.[15]

For many decades 'GABA' became synonymous with inhibitory neurotransmission. However, reports began to slowly emerge which made this generalisation feel increasingly uncomfortable. In the late 20th Century, inhibition within the CNS was a seemingly simple affair, with synaptic inputs from GABAergic and glycinergic neurons providing a calming counterbalance to excitation mediated by glutaminergic, cholinergic and serotoninergic influences. GABA receptor classification was a simple matter of ABC, and all three classes were believed to have an inhibitory mechanism of action.[16] This was because, under most experimental recording conditions, the intracellular Cl- concentration, or activity, is below the predicted electrochemical equilibrium potential for Cl-, or E_{Cl}. $GABA_A$ and $GABA_C$ class receptors were simply assumed to exert an inhibitory influence by mediating the GABA-gated *influx* of negatively charged Cl- ions across the cell membrane.

The equilibrium potential for Cl- (E_{Cl}) is determined by the transmembrane potential and the concentrations of Cl- on either side of the membrane. This is predicted by the Nernst equation in the form;

$$ECl = \frac{RT}{zF} . \ln\left(\frac{[Cl-]i}{[Cl-]o}\right)$$

---

[15] As K+ concentrations are usually very low outside the cell and high inside the cell, E_K tends to be very negative, whereas the opposite is true of Na+ channels, underlying the explosive depolarisations and repolarisations of action potentials.

[16] Bormann J. The 'ABC' of GABA receptors. Trends Pharmacol Sci. 2000 Jan; 21(1):16-9.

If the intracellular Cl- activity, or $[Cl^-]_i$, is higher than the predicted value for E_{Cl}, then $[Cl^-]_i$ is said to be *above* electrochemical equilibrium and thus Cl- will leave the cell if Cl- selective ion channels open, thereby exerting a depolarising influence upon the cell membrane potential and exciting the cell. In most neurons, however, $[Cl^-]_i$ is found to be below equilibrium and thus GABA and glycine are inhibitory.

Indeed, the role of Cl- in cellular electrochemistry is often overlooked, and some researchers even neglected to consider that the entry of Cl- ions into a neuron is electrochemically equivalent to the efflux of K+.[17]

Intracellular Cl- activity is, in essence, the great variable of electrochemistry, and whilst large electrochemical gradients for Na+ and Ca2+ entry are ubiquitously maintained, and the equilibrium (reversal) potential for K+ is usually near or beyond (i.e. more negative than) the observed resting membrane potential of the cell, commonly encountered values for E_{Cl} fall well within the operational range of membrane potentials observed in most excitable cells. Further, it is more economical in terms of energy for a cell to establish and maintain a $[Cl^-]_i$ which is above or below E_{Cl}, than it is to maintain the high Na+, K+ and Ca2+ gradients which are actively generated by the work of the Na+/K+ ATPase and the Ca2+ATPase. Thus, varying the intracellular activity of Cl- provides cells with a flexible alternative in the regulation of their excitability and in the generation of functional diversity within the nervous system.

Although it is well known that intracellular chloride is accumulated above electrochemical equilibrium within fluid and electrolyte secreting epithelia by means or a NaCl or Na+K+2Cl- cotransporter,[18] many reports have emerged to suggest that $[Cl^-]_i$ is also accumulated above equilibrium within a variety of cell types that functionally express $GABA_A$ receptors.

---

[17] Scott Nawy & David Copenhagen. Multiple classes of glutamate receptor on depolarising bipolar cells in retina. **Nature.** 1987 Jan 1-7;325(6099):56-8.
[18] O'Brien JA, Walters RJ, Valverde MA, Sepulveda FV. Regulatory volume increase after hypertonicity- or vasoactive-intestinal-peptide-induced cell-volume decrease in small-intestinal crypts is dependent on Na(+)-K(+)-2Cl- cotransport. Pflugers Arch. 1993 Apr;423(1-2):67-73.

Is GABA a bipolar neurotransmitter?

While a detailed discussion of the variations in E_{Cl} which are observed within the nervous system is highly complex and involved (please see Appendix L), we shall instead journey to the excitable interface between the endocrine (hormonal) and central nervous systems for a clearer illustration of the principle that the activation of membrane Cl- conductances by GABA may be excitatory.

$GABA_A$ receptors are widely expressed within a wide range of neuroendocrine cells, including insulin-secreting pancreatic β-cells, adrenal chromaffin cells and the model PC12 pheochromocytoma cancer cell line (see Appendix D). Within these neuroendocrine cells, GABA is known to evoke a membrane depolarisation, as [Cl-]i is maintained *above electrochemical equilibrium*. Thus, the activation of $GABA_A$ receptors would cause an efflux of Cl-, thereby depolarising the cell membrane.

Perhaps the most dramatic illustration of GABA serving as an excitatory transmitter is the presence of functional $GABA_A$ receptors on adrenal chromaffin cells. Given that GABA has been shown to evoke the release of adrenaline, there can perhaps be no more apt illustration of the role of Cl- (and of GABA) in cellular excitation?

As GABA has been established as a primary trigger for the release of catecholamines, all of its classical modulators, including the benzodiazepine (BDZ) family,[19] ethanol and neuroactive steroids, may also be expected to modulate adrenaline output, especially as I was able to establish the expression of a BDZ-sensitive alpha subunit within both the adrenal gland and PC12 cells (Appendix D). This would at first appear to give rise to a potential therapeutic paradox, given that BDZs are employed clinically as anxiolytics.

In summary, GABA can no longer be simplistically viewed as an inhibitory neurotransmitter, as GABA merely gates the flow of Cl- ions as the intracellular chloride activity directs. Perhaps most intriguing is the observation that the regulation of intracellular chloride activity

---

[19] Walters RJ, Hadley SH, Morris KD, Amin J. Benzodiazepines act on GABAA receptors via two distinct and separable mechanisms. Nat Neurosci. 2000 Dec; 3(12):1274-81.

provides cells with an efficient, plastic and very dynamic means of regulating their excitability.

In the next chapter, we shall introduce the complexities of single channel recording as we seek to identify which ion channels underlie the macroscopic currents observed within the small intestinal crypt.

Chapter Four

In search of the underlying single channels

Having established that the resting membrane potential of the small intestinal crypt is dominated by an *outwardly rectifying* K+ current and that there is also a Ca2+-activated K+ conductance present, I turned my attention to identifying which individual ion channel activities might account for the whole cell currents observed.

However, when one approaches the recording of single channels, our application of the patch-clamp technique changes dramatically. First, as we are seeking to identify the single channel, or unitary currents, which traverse the cell membrane through a single channel pore, we opt to use much finer electrodes with greater 'resistances' to increase the probability that we may isolate a single channel and record the currents that pass through it.

Second, we are no longer studying macroscopic currents and the aggregate behaviour of individual ion channels. In this mode, we are now studying the behaviour of a single ion channel. This not only allows us to study the gating kinetics of a single channel pore (*i.e. its pattern of openings and closures under a variety of conditions*), but also affords an insight into how the channel may be regulated.

Third, as the electrical resistance of the seal and the non-conductive regions of the membrane surrounding the individual ion channels are very large, we are no longer concerned with the issue of series resistance, as the system is now reduced to a simple circuit in which a small resistance (*i.e.* a single channel conductance) is in series with a battery (*i.e.* the transmembrane potential).

Fourth, we can study single ion channels in a variety of permutations. The two most common of these are the *'cell attached'* patch, in which the composition of the extracellular *milieu* (*i.e.* what we put in the patch electrode solution) is known and what bathes the cytoplasmic, or internal face of the membrane is an unknown. Despite its evident limitations, the cell attached patch has proven a powerful tool with which to investigate the physiological properties of single channels, as we can stimulate the cell and record the outcomes without any fear that we have disrupted intracellular processes. A second common single channel recording technique is to excise the membrane patch from the cell, the so-called *'inside-out'* mode, which enables us to control the solution bathing both faces of the isolated membrane patch. This affords us the opportunity to study both the ionic selectivity of the individual channels and their regulation by second messengers simply by changing the solution bathing the cytoplasmic face of the membrane patch.

However, as I pursued my investigation into the identity of the enigmatic single channels underlying the macroscopic currents, another unanticipated surprise awaited. The two most frequently observed single channels bore no overt similarity to the macroscopic currents recorded...

The first of these was a spontaneously active inwardly-rectifying single channel activity which was present in 29% of cell-attached patches (19/66 patches) obtained from the crypt *basolateral* membrane. A series of voltage steps was applied to the cell attached patch which are, of course, step changes that are *relative to the spontaneous membrane potential* of the crypt, as we can no longer command the voltage arising across the entirety of the cell membrane. A series of recordings from an inwardly-rectifying single channel are shown in the figure below with the arrows indicating the closed state of the channel. Note how the noise increases when the channel is open and that, even when the ion channel is closed, there is substantial background noise present.

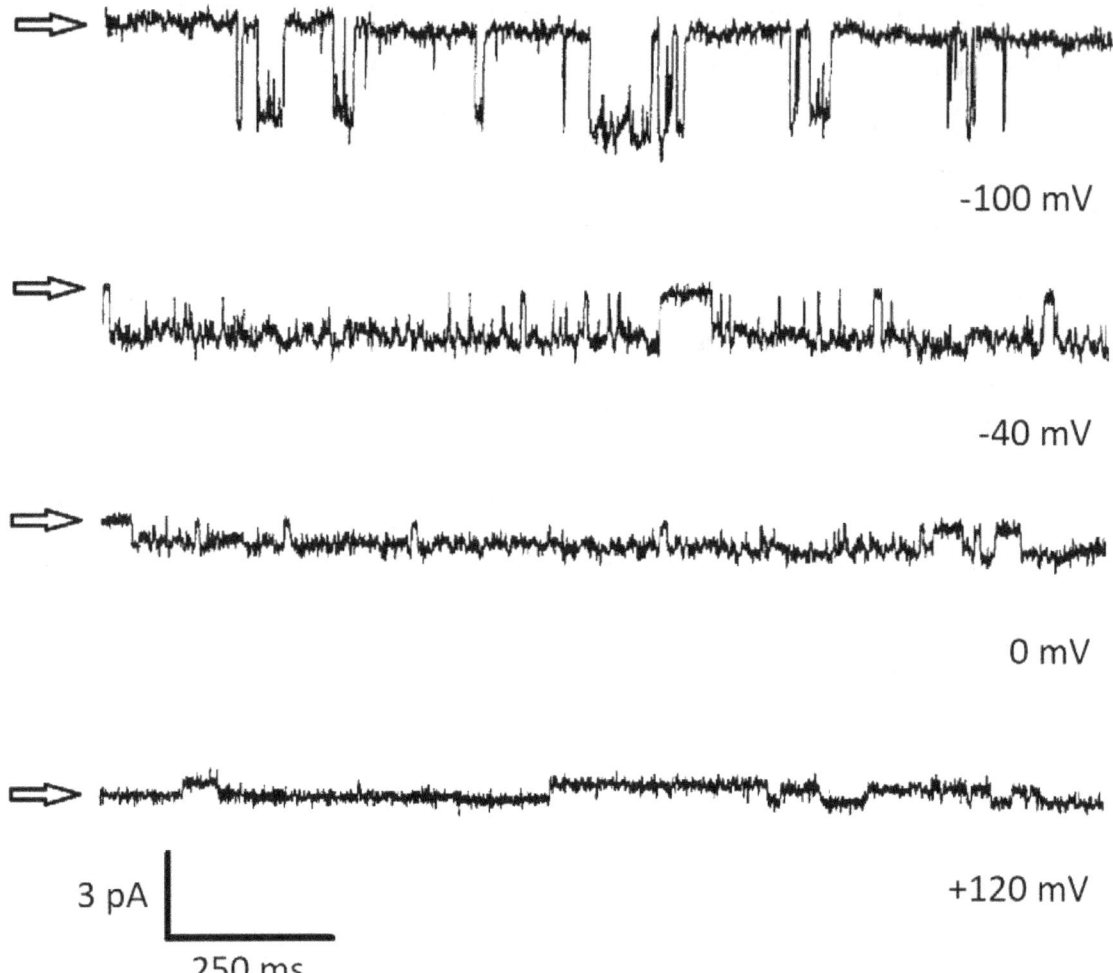

As we progressively change the holding potential of the cell attached patch from -100mV (*upper recording*) to +120mV (*bottom recording*) we observe changes in the pattern of single channel activity. First, the unitary conductance of the ion channel increases as the patch is hyperpolarised, and decreases as we depolarise the membrane patch, with a small outward current finally being observed at a holding potential of +120mV relative to the spontaneous membrane potential of the cell.

Calculating single channel conductance

Owing to the noise present within the single channel recording, we normally acquire and digitise each recording to produce a plot of frequency at each current value, otherwise referred to as a *'unitary current amplitude frequency distribution histogram'*. This allows us to estimate the mean unitary open current for any given potential from the peak value of the Gaussian distribution obtained. Presented below is a plot of frequency (*f*) vs. current (I in pA) for a patch containing multiple inwardly-rectifying channels at two holding potentials (-40mV, *left*, and -80mV, *right*). The vertical arrows indicate the channel closed state in which no channels are open within the patch. The unitary current *i* is calculated by determining the interpeak interval.

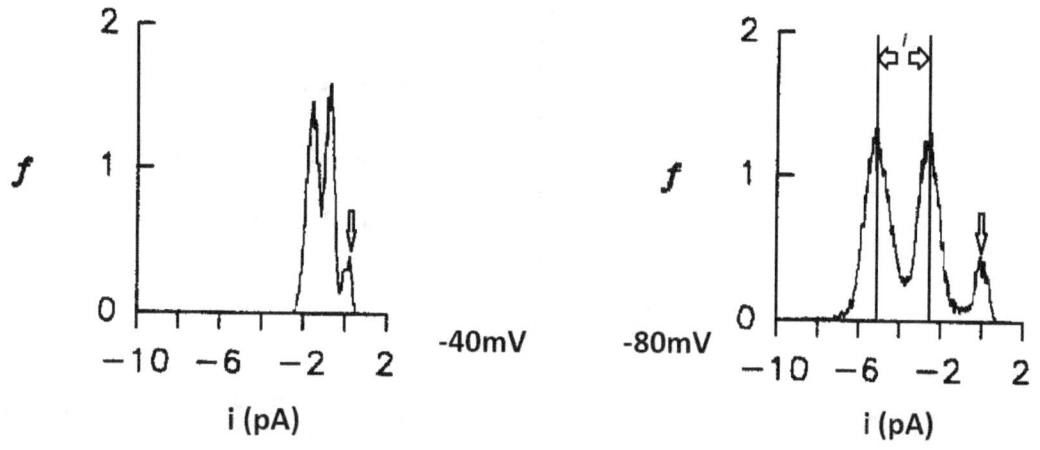

We then calculate the unitary conductance [20] (*g*) of a single channel as follows:

$$g = \frac{\Delta i}{\Delta V}$$

Where Δi is the change in unitary current for a given step change in holding potential (ΔV). In other words, we need to determine the *slope* of the single channel i-V relationship to calculate single channel conductance.

The resulting i-V relationship for the inwardly-rectifying single channel activity is shown below for the three recording conditions tested; namely in the cell attached mode (○); with the patch excised (inside-out/i.o.) with 140mM NaCl and 5mM KCl (●) bathing the intracellular face and again; inside-out (i.o.) with symmetrical 145mM KCl (□) on either side of the membrane patch.

---

[20] Conductance is the inverse of resistance, mathematically speaking.

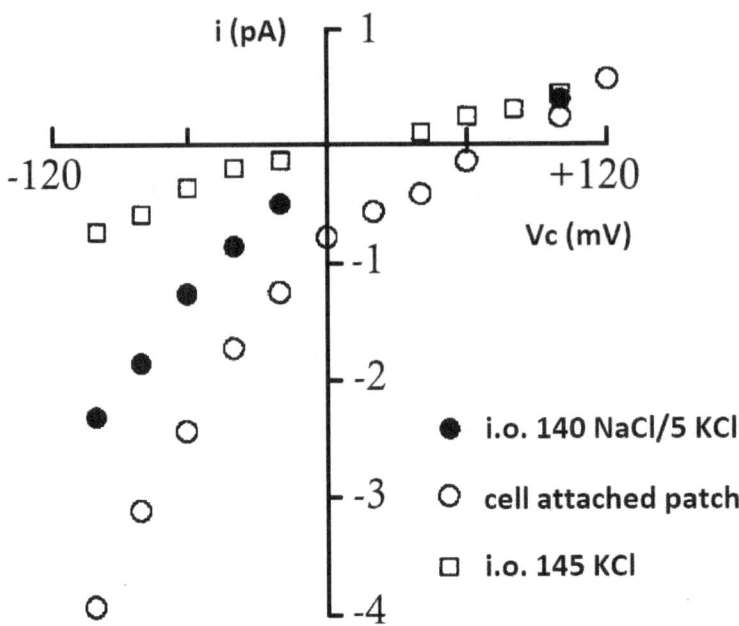

Even with a symmetrical 145mM concentration of KCl bathing both sides of the excised membrane patch (□), under which conditions we would expect a linear i-V relationship, the single channel conductance continues to exhibit an anomalous inward rectification, which is evidently more pronounced under cell attached conditions (○) and also where the intracellular face of the patch is bathed with a low K+ solution (●). The *shift* in reversal potential observed under recording conditions in which symmetrical 145mM K+ concentrations are present either side of the membrane to one where the intracellular K+ concentration is reduced from 145mM to 5mM (●) by the equimolar substitution of K+ with Na+ is consistent with a K+-selective channel activity. Under these *'unphysiological'* recording conditions (●), E_K would be +79mV. However, as this K+ channel activity exhibits an anomalous inward rectification, *i.e.* one which is not predicted by the GHK equation in the form;

$$i = g.V \frac{\left[C_o - C_i.\exp\left(\frac{FV}{RT}\right)\right]}{\left\{C_{sym}\left[1 - \exp\left(\frac{FV}{RT}\right)\right]\right\}}$$

where (i) is the unitary channel current, g is the single channel conductance, Co and Ci are, respectively, the ion concentrations bathing the extracellular and intracellular sides of the membrane patch, V is the command potential in the case of excised patches *(or the sum of the command potential and membrane potential in the case of cell-attached patches)*, R is the gas constant, T is the absolute temperature, and F is Faraday's constant.

As we cannot determine this channel's unitary conductance by means of a fit to the GHK equation, we must instead obtain the single channel conductance by means of fitting a tangent to the i-V relationship as presented above. Under cell attached recording conditions, this inwardly rectifying channel displayed a maximal slope conductance of 33 ± 0.8pS at a holding potential of -80mV and of approximately 16pS when the patch was held at the spontaneous membrane potential (i.e. no command potential was applied to the cell-attached patch).

An introduction to single channel kinetics

As this inwardly-rectifying K+ channel was observed to be spontaneously active in cell attached patches and only one channel open level was detected in 89% of such active patches, this enables us to perform a single channel kinetic analysis, as we logically cannot perform a kinetic analysis of the openings and closings of an ion channel where there is a mixed population of ion channels or more than one active channel within a patch.

To perform a kinetic analysis of the transitions of an ion channel between its open and closed states, we first need to digitally acquire a recording and then to parse the timeline for opening and closing events which are defined as transitions only if they cross the 50% open threshold as determined using a unitary current amplitude frequency distribution histogram *(as presented above)*. In the recording presented below, we denote the 50% transition threshold with the faint line *(the horizontal arrow denotes the channel's closed state)*.

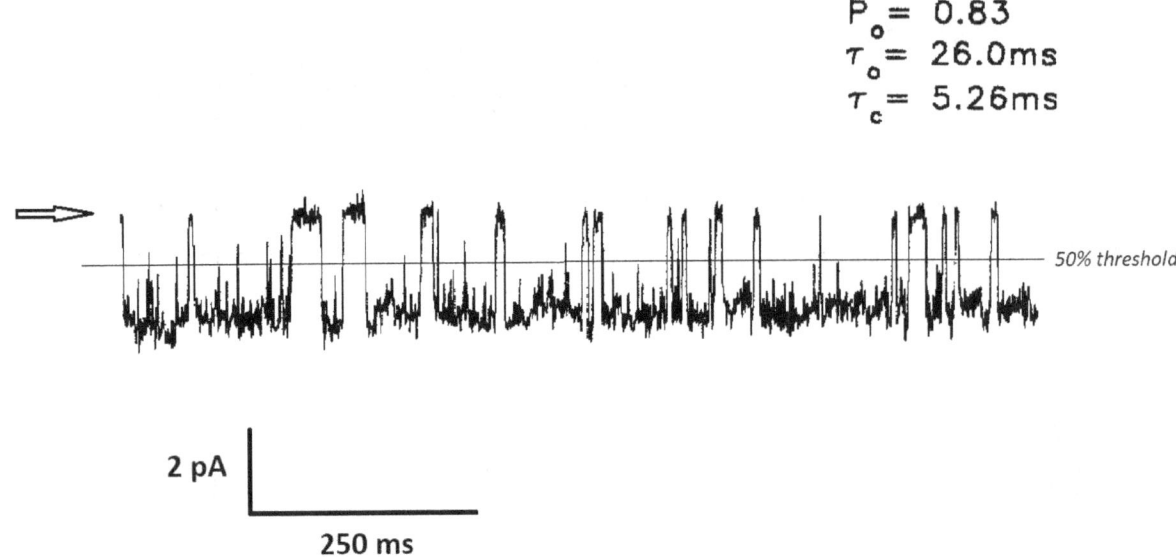

A kinetic analysis of such digitised single channel activity allows us to determine the precise *duration* of each channel closing and opening *event*. By this method, we can calculate the *mean* open time (τ_o), *mean* closed time (τ_c), and the single channel open probability (P_o), which is simply defined as the fraction of time a channel spends in its open state. Readily calculated, the determination of channel open probability is immensely useful in helping us to map single channel activity to macroscopic currents.

Noise and filtering considerations

When it comes to single channel recording, the importance of selecting the optimal acquisition and filtering frequencies is far too important to leave to the footnotes or appendices. Again, we shall approach this issue conceptually rather than mathematically.

Gating events within individual ion channels (*i.e.* transitions from the closed state to the open state, or *vice versa*) may occur at a much faster rate than the rate at which we sample, or digitally acquire any recording of single channel activity (4 kHz in these experiments). As a consequence, our ability to resolve high-frequency gating events may be lost.

The extent to which we *filter* a single channel recording is invariably a trade-off between our need to be able to resolve single channel events beyond the background noise, and our desire to maximise the integrity of the data. In other words, if we 'low pass' filter at 5kHz (*i.e.* we filter out all events with a frequency above 5kHz), then we will 'drown out' single channel transitions amid a sea of noise, especially those ion channels with a small unitary conductance. On the flipside, if we low pass filter much below 2kHz, then we lose the ability to record those single channel transitions which occur at higher frequencies.

For a low pass 8-pole Bessel filter, as used in these recordings, the performance of the filter is 'measured' in terms of its cut-off frequency, or f_c. Bessel filters are designed to eliminate high-frequency background noise in order to enable events of a lower frequency to be recorded, but their use does carry a significant limitation – a limit to the fastest events that we are able to record. The other limitation is, of course, the rate of acquisition, or sampling rate, which, for the purposes of these single channel recordings, was 4kHz.

The corner frequency of the Bessel filter, or f_c, is most usefully described in terms of its rise time t_r, which is given by:

$$tr = \frac{0.3321}{fc}$$

which, for a low pass Bessel filter with a setting of 1kHz, yields a rise time of 0.322ms. Thus, the sampling rate should always be at least 4 times, and ideally more than 10 times greater than the fastest event frequency being recorded. Here, as we are recording using a low pass Bessel filter with a setting of 1kHz and sampling at 4kHz,[21] we are at the limit of acceptable recording conditions although, as can be seen from the recording above, we are inevitably losing some of the faster transitions from the closed to the open state, and *vice versa*.

While this will not substantially distort the measurement of single channel open probability (P_o), we will however greatly distort measurements of mean open time (τ_o) and mean closed time (τ_c), which serve as key parameters in studying the gating kinetics of single channels.

If we observe the behaviour of the inwardly-rectifying K+ channel more closely, we can observe that the channel's open probability P_o increases as the membrane patch is

---

[21] Typical of the technology of the early nineties employed here.

'depolarised' from a resting membrane potential of around -60mV. This is reflected in the single channel recordings *(shown above)* and in the measurement of P_o from the resulting *unitary current amplitude frequency distribution histograms*.

P_o is simply given by;

$$Po = \text{time spent in open state/total recording time}$$

A plot of single channel open probability (P_o) as a function of holding potential (V_c) for the inwardly-rectifying K+ channel is shown below.[22] Note that the open probability of the inward rectifier is strongly voltage-dependent.

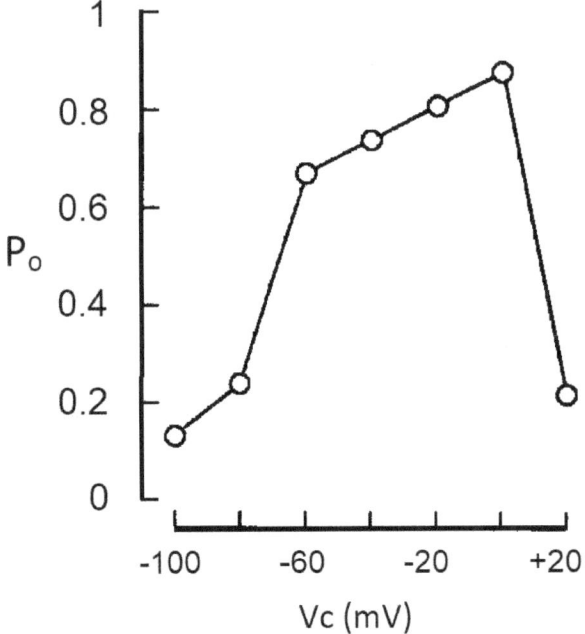

Patches containing inwardly-rectifying single channel activity were excised to ascertain the dependence of channel activity upon the presence of cytosolic Ca2+ *(see recordings below)*. The three recordings shown are from an excised ('inside-out') patch at a holding potential of -100mV in the presence of 1µM Ca2+ *(upper trace)*, 10nM Ca2+ *(central trace)*, and 1mM Ca2+ *(lower trace)*, respectively.

---

[22] Kinetic analysis was performed on cell attached recordings of 30s duration.

While these experiments are inherently *unphysiological* (*e.g.* the high extracellular concentration of K+ within the patch pipette; applying a range of holding potentials that the cell membrane would never realise; and the use of high cytosolic Na+ and Ca2+ concentrations in determining ionic selectivity), we are nonetheless able to generate a biophysical 'fingerprint' for the inward rectifier in terms of its slope conductance, selectivity for K+ ions, and the dependence of its activity upon voltage and the presence of cytosolic Ca2+.

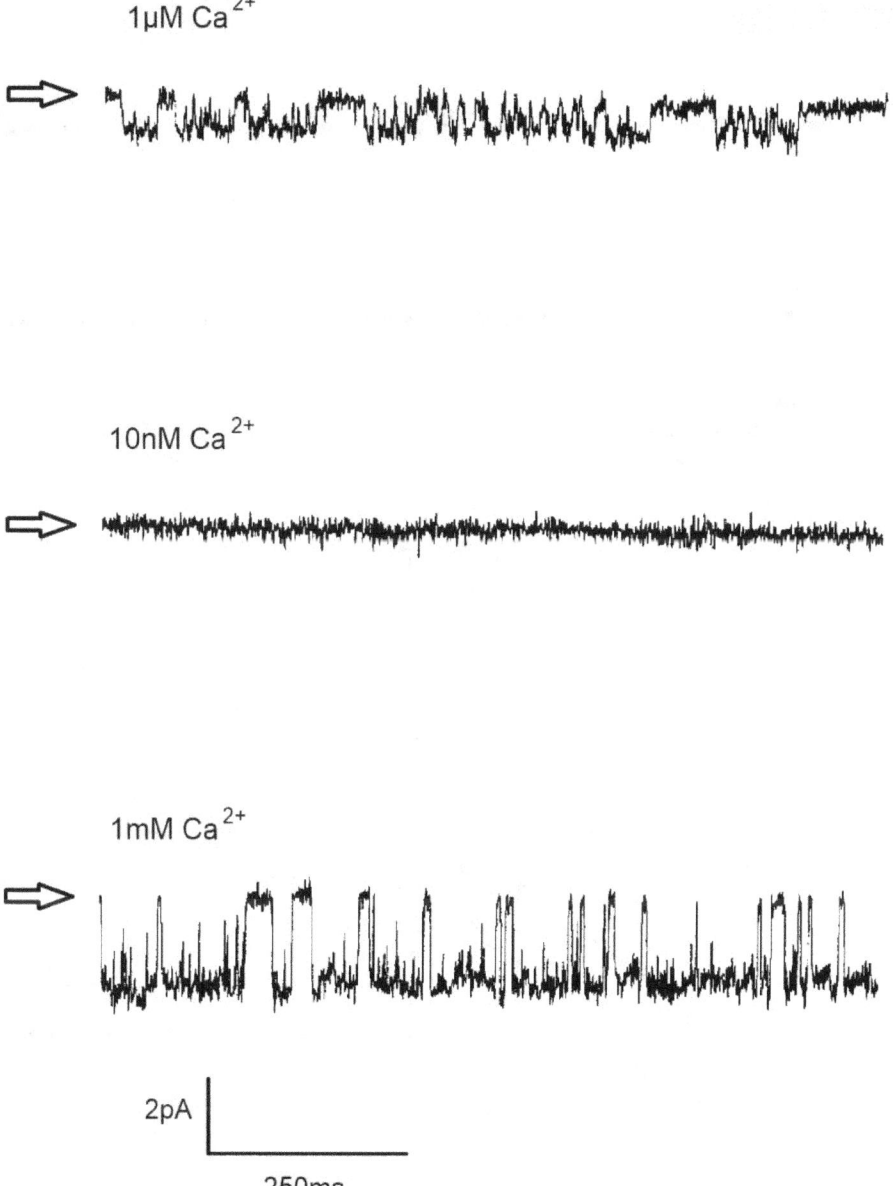

A second single channel activity within the crypt basolateral membrane

A second type of single channel activity was observed in only 6% of cell attached patches (4/66 patches) obtained from the basolateral membrane, which was characterised by brief channel openings that became more frequent as the membrane patch was hyperpolarised.

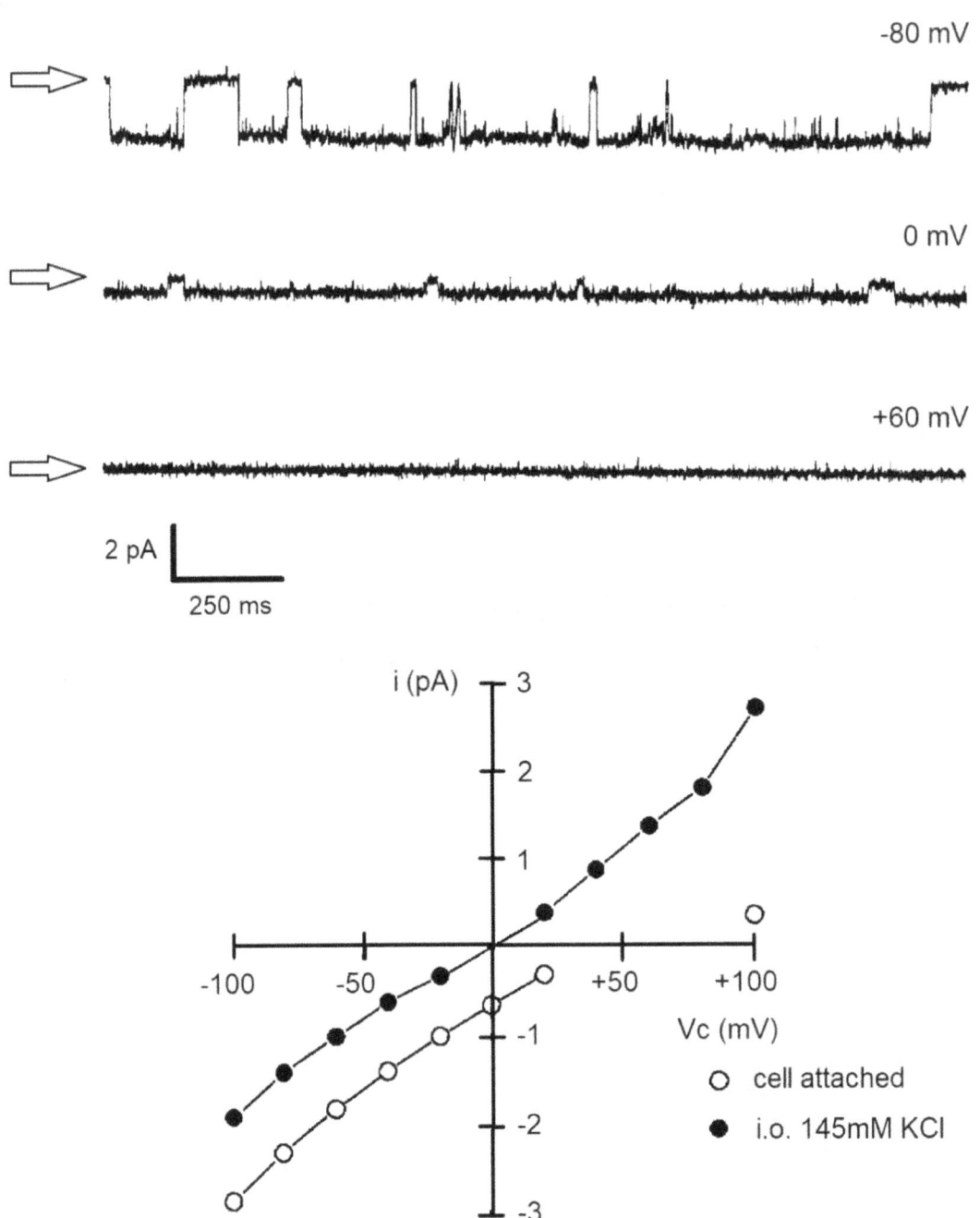

An analysis of the resulting i-V relationship in cell attached patches gave a slope conductance of around 18pS at 0mV and of 26-27pS at a holding potential of -80mV with a reversal potential of +50mV. This is also consistent with a selectivity to cations over anions, as [Cl-]$_i$ would need to be between 150 and 180mM to account for such a reversal potential. Note also that the i-V relationship reverses at 0mV in symmetrical 145mM KCl, indicating that this ion channel is also permeable to K+ ions.

When patches containing this channel activity were excised into solutions containing 1mM [Ca2+]i, the level of channel activity increased dramatically and, in those cell attached patches in which only a single channel had appeared to be present, multiple channel open levels were subsequently observed upon excision, suggesting that this channel is found in clusters.

The ionic composition of the medium bathing the cytosolic face of the excised patches was varied to determine the selectivity of this channel. When the cytosolic face of the patch was exposed to 145mM (symmetrical) KCl (in the presence of 1mM [Ca2+]i), the channel exhibited a linear i-V relation (*see above*), which reversed at 0mV with a slope conductance of 17-18pS. When the cytosolic face of the patch was then bathed with a solution that was high in Na+ or NaGluconate, the channel i-V relationship still reversed at 0mV with a linear slope conductance of 18-20pS (*see below*). However, when the cytosolic face was bathed with 40mM KCl/105mM NaCl, the channel i-V relationship exhibited a slight outward rectification which could be described by the GHK equation (*i.e.* the channel exhibits Goldman rectification) in the following form;

$$i = g.V \frac{\left[C_o - C_i.\exp\left(\frac{FV}{RT}\right)\right]}{\left\{C_{sym}\left[1 - \exp\left(\frac{FV}{RT}\right)\right]\right\}}$$

The resulting fit to the GHK equation yields a slope conductance of 18pS, a reversal potential of -10mV and a permeability ratio (PNa/PK) of 1.7:1. Therefore, this channel is classified as a non-selective cation channel, albeit one which is slightly more permeable to Na+ ions than to K+ ions.

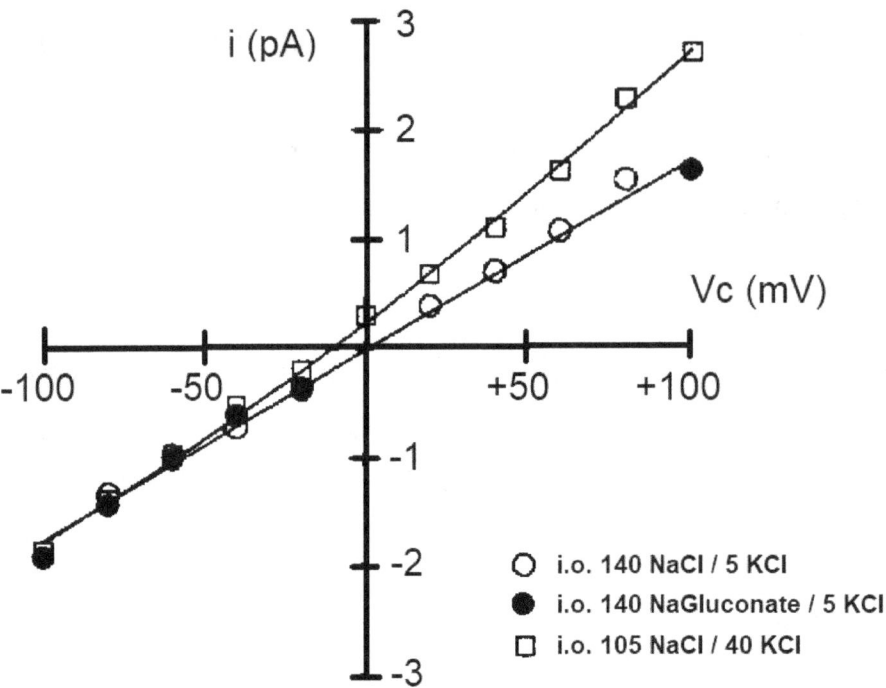

Experiments were then performed to establish the Ca2+-dependence of this cation non-selective channel activity in excised inside-out patches. When the cytosolic free Ca2+ was buffered to less than 10nM, channel activity was almost completely abolished at all holding potentials, although activity was readily restored upon the reintroduction of 1μM Ca2+ to the medium bathing the cytosolic face. When we plot the channel open probability for multiple channels within a patch (nPo) as a function of holding potential, nPo appeared to increase markedly as the patch was hyperpolarised in the presence of 1μM cytosolic Ca2+.

Previously, we defined channel open-state probability (Po) as the fraction of the total time that the channel occupies the open state, which is computed using the detection threshold method and the sum of the duration of channel open times derived. As in this instance, when patches contain more than one channel, the detection threshold was reset for each successive single channel level to determine the time intervals when 1, 2, 3,...n channels were open simultaneously. Thus, channel open probability for a patch containing more than one channel (nPo) was calculated as Po (level 1) + Po (level 2) ... + Po (level n). If a patch contains a known number of channel open levels, single channel Po may simply be determined by dividing nPo by n.

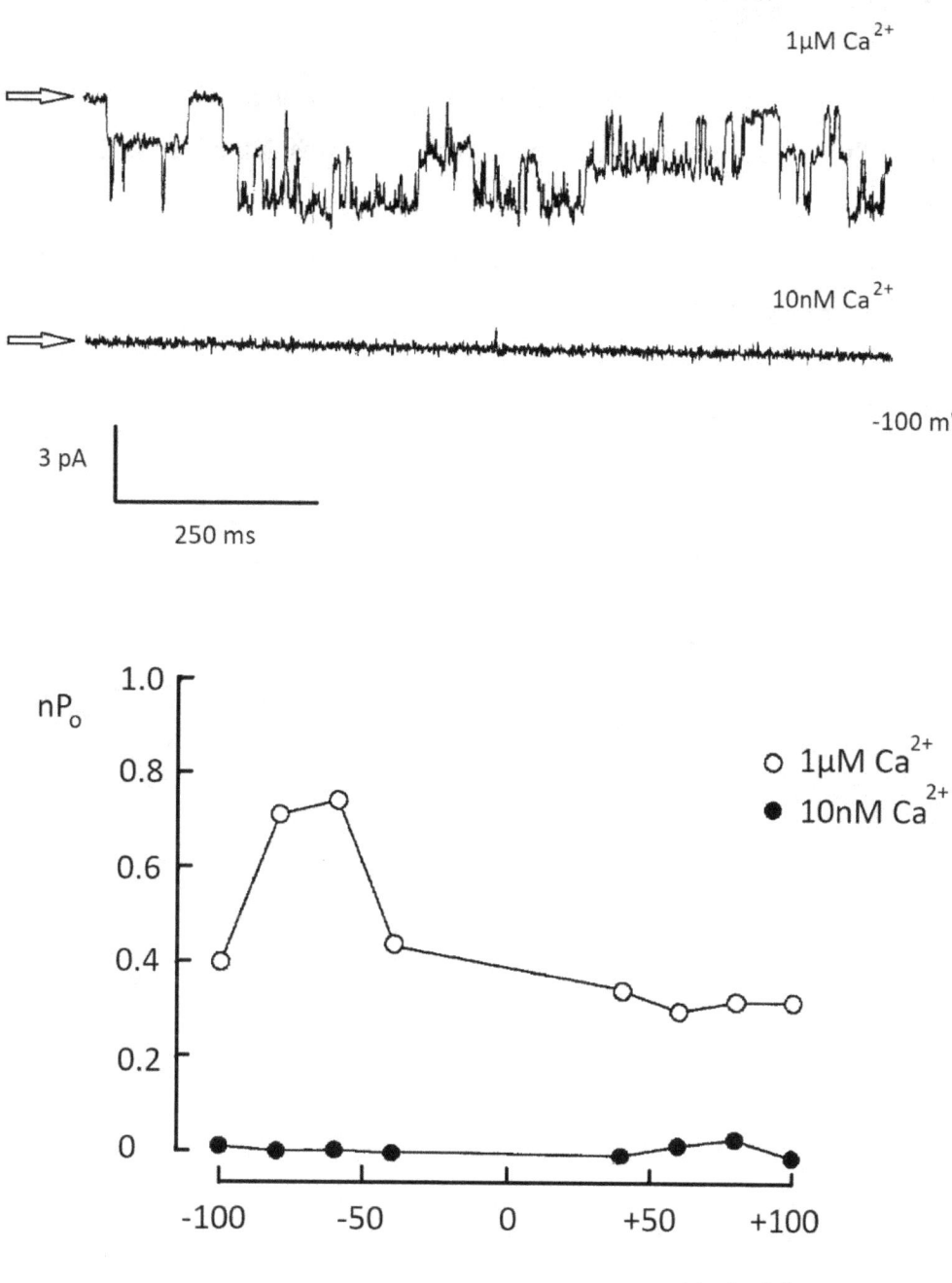

We have now established that this single channel activity belongs to the ubiquitous family of Ca2+-activated cation non-selective channels, or CAN channels. However, other than showing that this cation non-selective channel is Ca2+-dependent in its activity, we have yet to demonstrate that it is activated under more physiological conditions by a drug that triggers the release of intracellular Ca2+.

In our continuing search for the single channel activity underlying the Ca2+-activated K+ channel of the crypt basolateral membrane, carbachol was thus added to the medium bathing the crypts while recording in the cell attached mode of the patch clamp technique. As we can see below, the addition of 100μM carbachol evoked large increases in single channel activity in patches which were previously quiescent (i.e. silent).

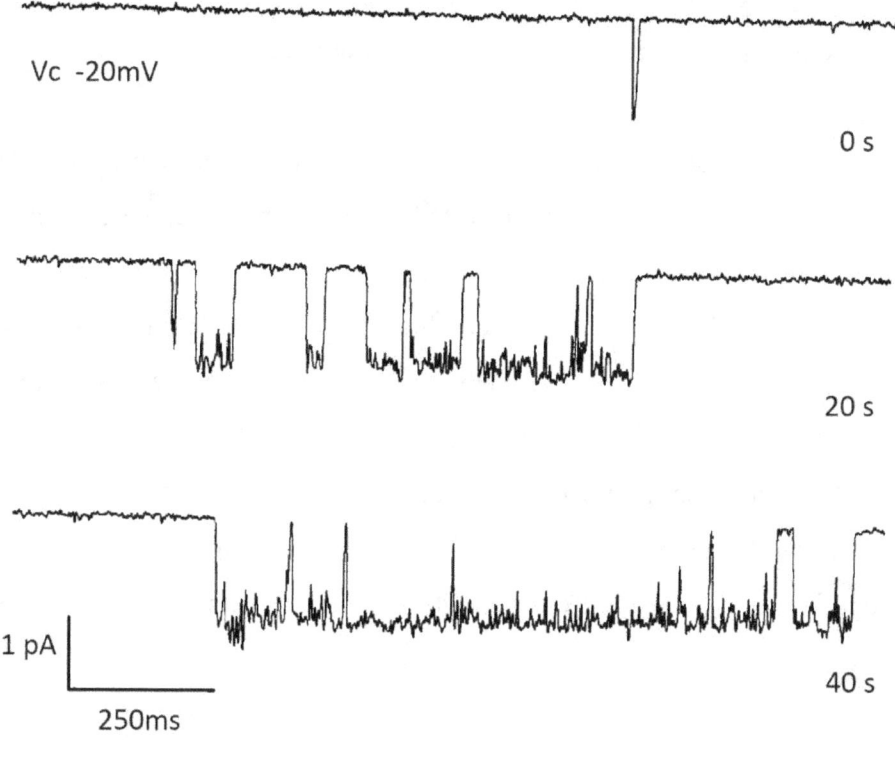

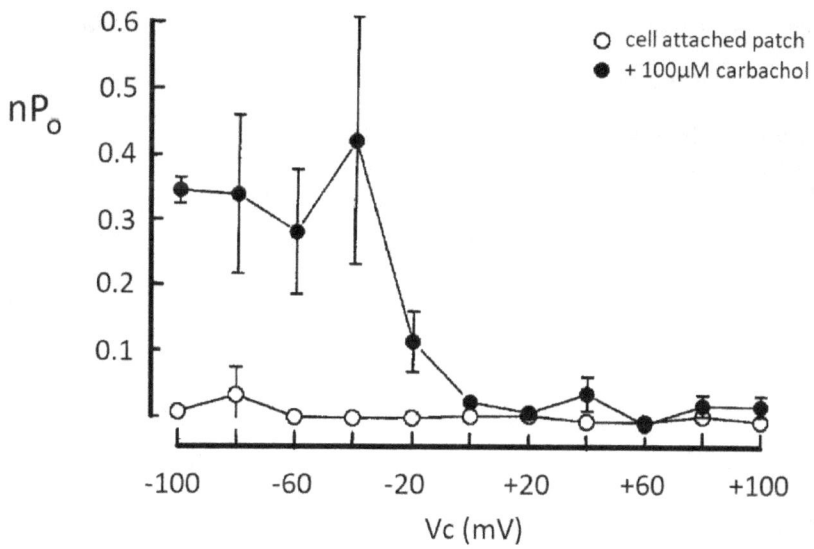

The recording presented above shows a recording of single channel activity in the cell attached mode at a holding potential of -20mV at the time intervals of 0 s, 20 s, and 40 s after the addition of carbachol. The graph immediately beneath it shows mean values of nP_o (± SEM) as a function of holding potential from three separate experiments across a range of holding potentials in the presence and absence of carbachol. Again, this carbachol-evoked increase in single channel activity, which was observed in 64% (14/22) of cell attached patches, was most notable at hyperpolarised holding potentials.

To recap, we have noted the presence of an inwardly-rectifying K+ channel with a unitary conductance of 32-34pS which is activated by membrane depolarisation and is Ca2+-dependent in its activity. We have also identified a Ca2+-activated cation non-selective (CAN) channel activity with a unitary conductance of 18pS. Although this CAN channel activity appears to be activated by carbachol in cell attached patches, its greater permeability to Na+ ions than to K+ ions makes it an awkward candidate to explain the Ca2+-activated K+ channel activity identified from whole cell conductance measurements. Thus, we need to refine our search and to be more discerning in our choice of experimental conditions if we are to identify a candidate single channel which might explain either the Ca2+-activated K+ conductance or the resting membrane K+ current.

Engineering ionic gradients in single channel recording

In the following series of experiments, the pipette solution was modified to allow me to identify any K+-selective single channel activity arising in cell attached patches without the need for the patch to be excised for ion substitution experiments. This was achieved by the use of pipette solutions which contained low concentrations of KCl and high concentrations of NaGluconate. The rationale for this was that E_K would thus be very negative (even though $[K+]_i$ is unknown), while E_{Cl} and E_{Na} would be very positive. *Thus, any outward currents appearing at physiological membrane potentials must be carried by K+ ions.*

To probe for Ca2+-activated K+ channel activity, cell attached patches were obtained and 100μM carbachol was again applied to the bathing solution. Small outward currents, indicative of single channel activity, were induced within seconds of carbachol addition at the spontaneous membrane potential and the amplitude of the single channel current increased as the membrane patch was depolarised with an extrapolated reversal potential that was around 50mV more hyperpolarised than the spontaneous membrane potential. This small conductance channel activity, shown below, was observed in 6/18 (33%) of cell-attached patches after stimulation with carbachol.

control

Em + 0mV

100µM carbachol + 20s

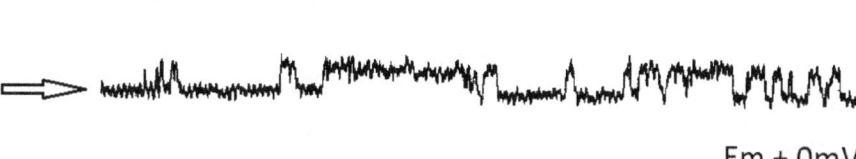

Em + 0mV

100µM carbachol + 80s

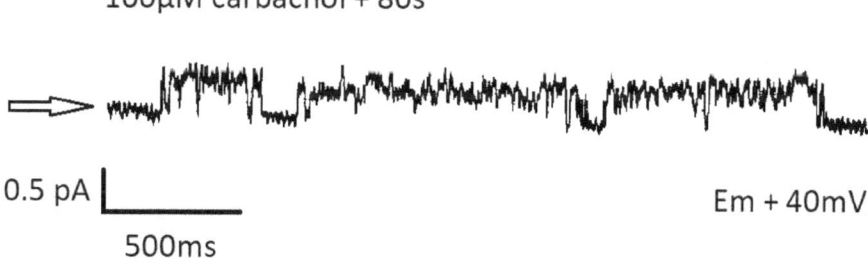

0.5 pA

500ms

Em + 40mV

The figure above shows the effect of carbachol upon single channel activity recorded in cell attached patches with a low KCl pipette solution. The upper trace shows a representative recording from a cell attached patch which was held continuously at the spontaneous membrane potential for 1 minute prior to agonist addition. The central trace shows a recording of channel activity at the spontaneous membrane potential 20 s after the addition of 100µM carbachol to the bathing medium, and the lower trace shows a recording of channel activity in the presence of agonist at a holding potential of +40mV, some 80s after the addition of carbachol. Recordings were low pass filtered at 0.5kHz to be able to resolve single channel events.[23]

The carbachol-induced channel activity appeared within 3 to 5 seconds of agonist application in these patches, a time-course consistent which is consistent with the hyperpolarisations of Em observed in the current clamp mode. No large conductance K+-selective single channel activity was observed in response to the addition of carbachol in any of the 18 cell attached patches tested. When I attempted to excise the membrane patches containing this small conductance K+ channel, all activity was lost and no outward K+ currents were observed in any of the 18 recordings with low 'extracellular' KCl concentrations.

---

[23] Such tiny single channel currents could not be resolved using a low pass filter setting at 1-2kHz.

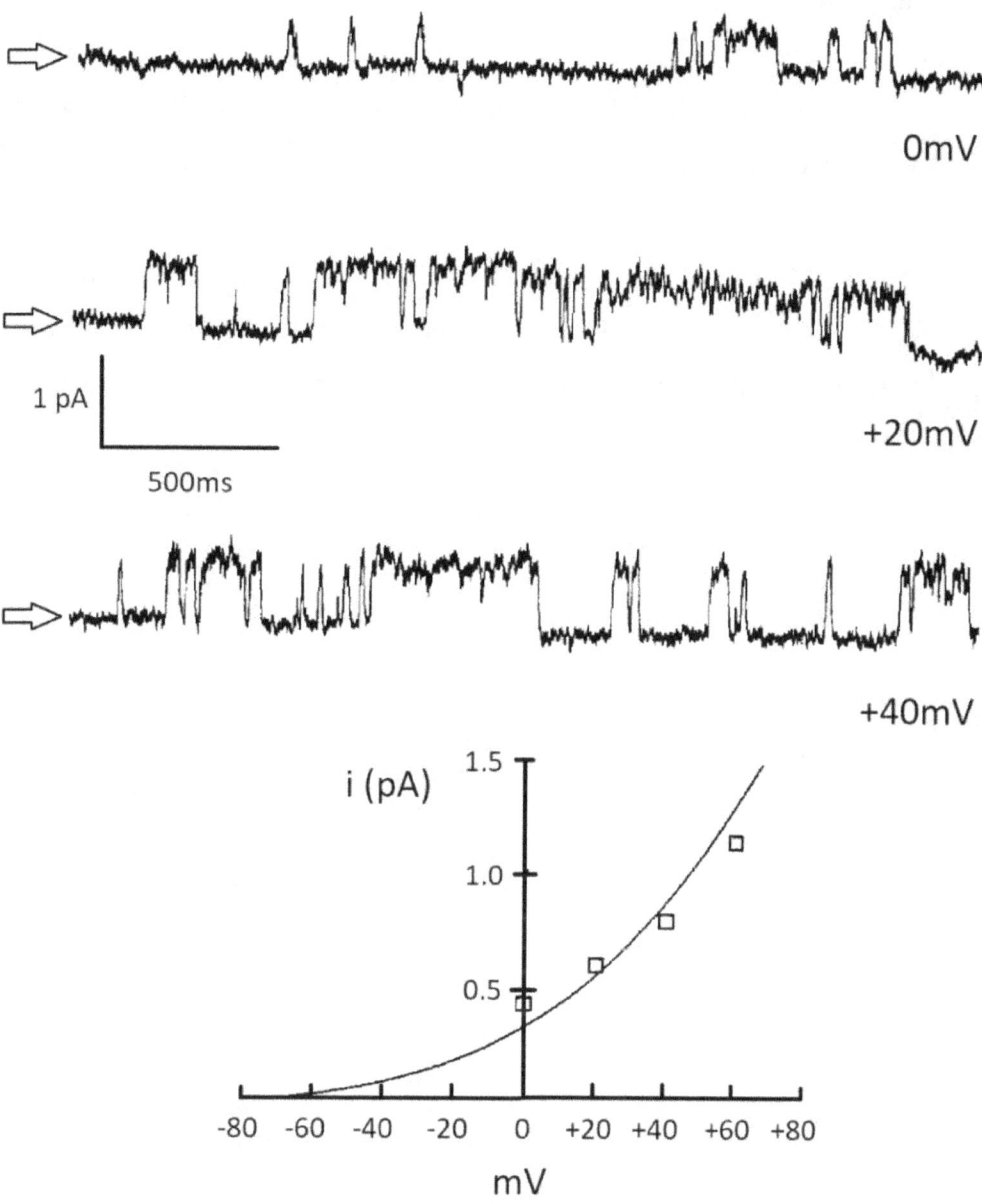

It is interesting to note that outward K+ currents were also activated in 3 cell attached patches in response to a 20mV step depolarisation of the membrane, without the need for a prior application of carbachol. An example of this activity is shown in the recording above. The resulting i-V relationship was fitted to the GHK equation, assuming a membrane potential of -45mV and [K+]$_i$ and [Na+]$_i$ activities of 125 and 20 mM respectively. This fit to the GHK equation indicated a reversal potential of -70mV, a slope conductance of 3.7pS and a PNa/PK of less than 0.05, suggesting that this small conductance channel is highly selective for K+. However, this small conductance K+ channel activity was not observed at hyperpolarised potentials. This depolarisation-activated K+ channel yielded i-V relationships with mean slope conductances of 2.0, 2.8 and 3.7pS (2.8 ± 0.5pS) and corresponding reversal potentials between -70 and -75mV more negative than the spontaneous membrane potential. Thus, these the properties of these low conductance K+ channels are consistent with the carbachol-activated K+ conductance.

Thus, I was able to identify the single channel activity underlying the Ca^{2+}-activated K+ conductance, but what of the K+ channel activity which correlates to the outwardly-rectifying K+ conductance of the small intestinal crypt basolateral membrane?

Chapter Five

Recording the activity of very small ion channels

As we shall soon learn, not all the openings and closures of ion channels present within the cell membrane may be directly observed using the patch clamp technique. Even for single channel recordings with exceptionally low background noise, which was reduced to less than 0.1pA in these recordings, some ion channel events may be too small to resolve directly.

One approach that was developed to address this issue is *noise analysis*, but we shall not cover that in this work.[24] In this chapter, we shall take an entirely different approach, although a combination of both methods may be optimal.

Previous whole cell recordings indicated that the resting membrane potential of small intestinal crypts is dominated by an outwardly-rectifying K+ conductance. However, single channel recordings presented in the previous chapter failed to identify any prospective candidates for the underlying channel activity. These observations raised the intriguing possibility that such a conductance may be too small to resolve at the single channel level.

I thus performed ion substitution experiments upon excised inside-out basolateral membrane patches using a pipette solution which contained 145mM Na gluconate, 1.3mM $CaCl_2$, 0.5mM $MgCl_2$ and 1mM K gluconate which had been designed to isolate outward K+ currents.

When basolateral membrane patches were excised and the cytosolic face of the patch pipette was bathed with a solution containing 135mM Na+, 10mM K+, 148mM Cl-, 0.5mM Mg2+ and 36nM Ca2+, we can calculate specific the equilibrium potentials for each monovalent ion corresponding to -59mV for E_K, +1.8mV for E_{Na}, and +95.5mV for E_{Cl} according to the Nernst equation:

---

[24] For a more detailed discussion of *noise analysis* please refer to **Ion Channels**, by David Aidley & Peter Stanfield, Cambridge University Press.

$$Erev[X+] = \frac{RT}{zF} \cdot \log_e \left(\frac{[X+]o}{[X+]i}\right)$$

where RT/zF = 25.67 at 25° C.

During a continuous recording of patch current (Ip) from an excised inside-out patch which was initially perfused with a cytosolic solution containing 0.1mM KCl and 145mM NaCl ([Ca2+]i was buffered to 36nM) a particularly unusual phenomenon was observed. When the concentration of K+ bathing the cytosolic face was increased to 145mM by equimolar replacement of NaCl with KCl, with the cytosolic [Ca2+]i initially buffered to 1µM and then to 36nM with EDTA, there was a sharp increase in outward current without any overt changes in single channel activity. All such changes in evoked Ip upon substitution of K+ for Na+ were readily and rapidly reversible and, when 135mM NaCl was substituted for 135mM N-methyl D-glucamine chloride (NMDG+), an impermeant cation (while [K+]i and [Ca2+]i were maintained at 10mM and 36nM respectively), a sharp decrease in Ip was observed, which was again readily reversible. Mean changes in Ip for each ion substitution manoeuvre are given in (Appendix E, table 2).[25]

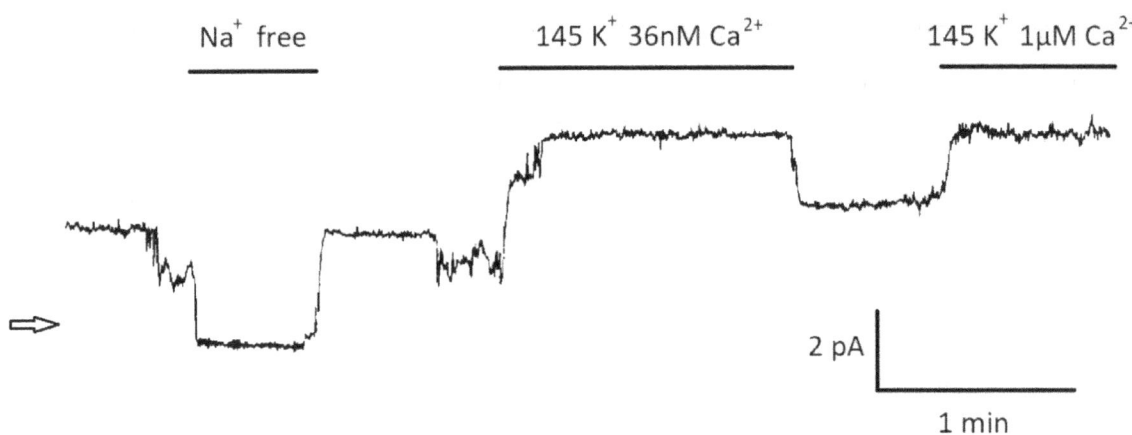

The continuous voltage clamp recording shown above indicates the cationic permeability of an excised inside-out membrane obtained from the crypt basolateral membrane (see Appendix E, table 1 for the composition of solutions). During the periods indicated by the

---

[25] The bursts of inward current observed in this recording likely reflect a separate and distinct channel activity.

horizontal bars, the solution bathing the intracellular face of the membrane patch was changed from a control solution of composition 145mM NaCl, 0.1mM KCl (with free [Ca2+]i buffered to 36nM with EGTA), first to a solution in which there was an equimolar substitution of 135mM N-Methyl D-Glucamine for NaCl, and then, after a return to the control solution, to a solution containing 145 KCl (by equimolar replacement for NaCl) with either an unchanged (36nM) or elevated free [Ca2+]i (1μM). The arrow indicates the zero-current level across the membrane patch and the horizontal bars the duration of solution changes.

Having identified a current across the membrane patch that appeared not to be carried by a leak current *(which are by definition non-selective and always reverse at 0mV)*, a series of experiments were then performed to determine the selectivity of this patch permeability pathway by recording the currents elicited by a series of voltage steps. The cytosolic face of the patch was held at -50mV, a value typical of the crypt resting membrane potential as measured by the perforated-patch technique, and a series of voltage pulses of 3 seconds in duration were applied to the pipette in 40mV increments from -80mV to 80mV.

The figure presented on the following page shows the current families that were elicited by a series of command voltages across an excised membrane patch in which the cationic composition of the cytosolic medium was changed *from* 10mM K+, 135mM Na+ *to* either 145mM K+ or 135mM NMG+ and 10mM K+. In these experiments [Ca2+]i was buffered to 36nM, a value typical of the resting cytosolic Ca2+ level reported in quiescent cells. The average current at each holding potential was determined and the resulting current-voltage (I-V) relationship derived is shown below.[26] In this experiment, the substitution of 135mM KCl for 135mM NaCl resulted in a shift in the I-V reversal potential from -12 to -26mV, which was associated with an increase in the magnitude of outward current at depolarised potentials and a greater outward rectification of the I-V relation. This is consistent with a K+-selective current with a unitary conductance of below 1pS.[27]

---

[26] The 'spikes' observed are capacitive currents arising from the charging and discharging of the glass electrode and could not be completely eliminated by capacitance neutralisation or by coating the electrodes by Sylgard.
[27] Arrows indicate zero-current levels.

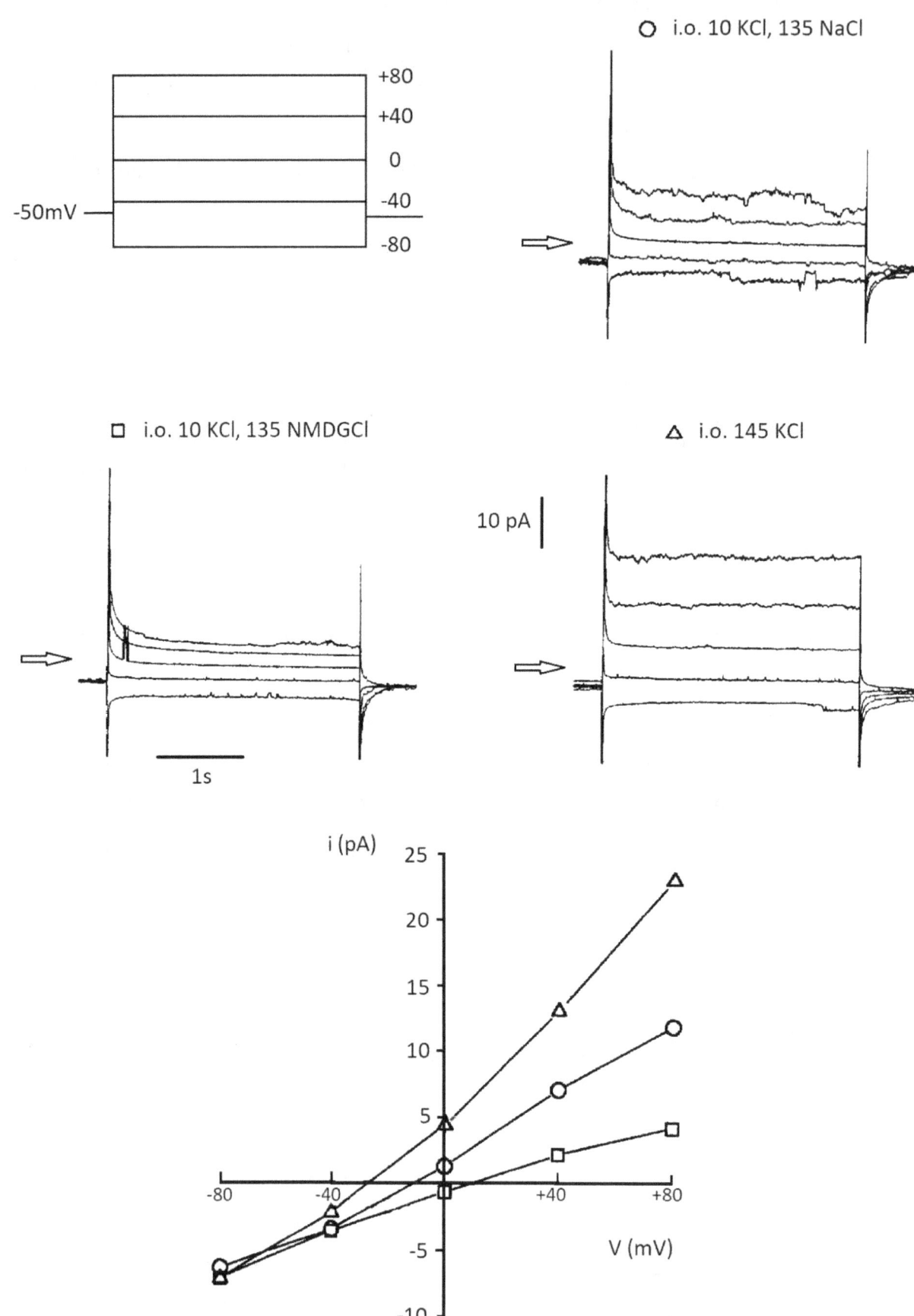

Substitution of 135mM NaCl with 135mM NMDGCl at the cytosolic face of the membrane patch resulted in a decrease in the outward current which was associated with an inward rectification of the I-V relation and a shift in the I-V reversal to +8mV (this data is summarised in Appendix E).

Further experiments were performed to determine the effect of increasing cytosolic free Ca^{2+} from 36nM to 1μM upon the size of the outward currents elicited by the same family of voltage steps. No significant difference in the magnitude of the outward currents evoked was observed *(not shown)*.

However, a time-dependent activation of outward current at more depolarised potentials was evident in most (13/18, or 72%) excised patches *(as show below)*, although this time-dependent activation did not appear to be dependent upon $[Ca^{2+}]_i$. However, in 5/18 (28%) of patches the outward currents were linear or slowly inactivating at depolarised holding potentials, as in the recording above.

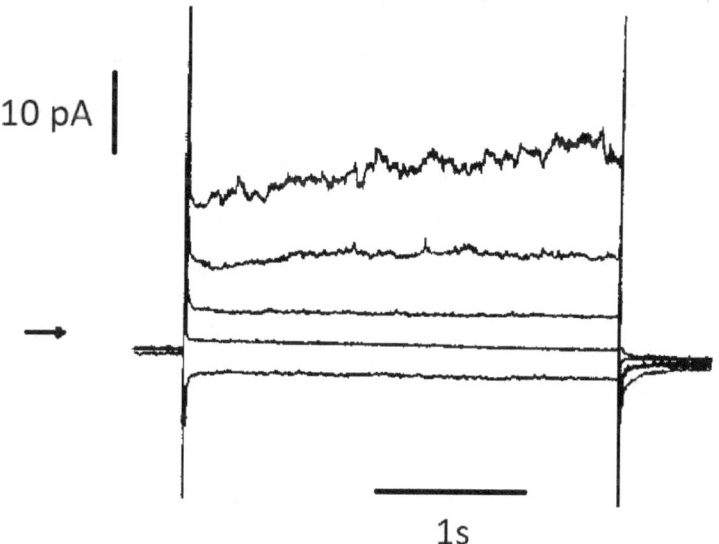

The increase in outward current and the negative shift in the reversal potential associated with substitution of Na^+ with K^+ are consistent with the presence of a permeability pathway that is selective for K^+ over Na^+ ions. However, the large decrease in outward current and the positive shift in the reversal potential that occurs upon substitution of Na^+ with NMG^+

also suggests a substantial permeability to Na+ ions, as the [K+]i was not altered in these experiments. The mean reversal potential obtained with cytosolic concentrations of 10mM KCl and 135mM NaCl was -17mV, from which value a PK/PNa of 18:1 can be calculated using the Goldman-Hodgkin-Katz (GHK) equation in the form;

$$EREV = \frac{RT}{zF}.\ln\left(\frac{PNa.[Na+]o + PK.[K+]o}{PNa.[Na+]i + PK.[K+]i}\right)$$

assuming that no leak permeability component was present.

However, this assumption is unlikely to be entirely valid and, therefore, such a value is probably an underestimate of the true PK/PNa ratio. Following an increase in cytosolic KCl from 10 to 145 mM, the mean reversal shifted to -27mV, from which a PK/PNa of 2.9 could be calculated, considerably lower than the value obtained with 10 K+ in the bathing medium.

If we assume that a PK/PNa of 2.9, a PNa/PNMG of 2.44 and a PK/PNMG of 9.1 may be calculated from the mean reversal potentials shown in the figure presented above, then such a discrepancy between the apparent permeability ratios obtained with 145mM KCl and 10 mM KCl could not be accounted for by a junction potential or by a deterioration in the integrity of the seal, as the changes in whole-patch current and reversal potential were completely reversible. One proposal to explain this phenomenon is that the selectivity of this conductance pathway may be regulated in some manner by the cytosolic Na+ and/or K+ activities, with the selectivity of this pathway to K+ appearing to decrease as the intracellular [K+]i is increased and the intracellular [Na+]i is decreased.

Irrespective of the conundrum of shifting permeability ratios and a selective cationic current that is too small to study in the form of discrete channel openings and closures, we have nonetheless identified a Ca2+-independent outwardly rectifying K+ current within the basolateral membrane which is activated by membrane depolarisation and is overtly consistent with the whole cell currents observed in the small intestinal crypt. It would be interesting to learn if this current, which appears to be present in all membrane patches, is also sensitive to quinine.   It would be even more intriguing if this current was sensitive to *ouabain*, a highly specific inhibitor of the Na+/K+-ATPase…

Chapter Six

The anatomy of action potentials

The classical era of electrophysiology dates back to the early 20th Century when the action potentials arising within the squid giant axon became a great scientific curiosity. From this formative work comes our fundamental understanding of those regenerative currents which arise from the explosive opening and rapid closure of excitable ion channels in concert which make possible the phenomenon of the action potential and the rapid transmission of information within the central nervous system. Early investigators were able to develop electrical recording devices, known as amplifiers, which could monitor the repolarising waves of current that propagated along the squid giant axon which is, in effect, a cable of cytoplasm ensheathed by an insulating membrane.

All of our classical understanding of equilibrium potentials, ion channel selectivity, and cable theory derive from this work, and I received my own introduction to the fascinating world of action potentials when I was asked to study them within the retinal ganglion cell, a class of neuron that relays light signals from the retina to the visual cortex.

However, before we address the question as to whether retinal ganglion cells are *anisopotential*, we need to revisit the fundamental nature of action potentials and the regenerative currents that underlie them.

The anatomy of an action potential

Initially, one of the most confusing aspects of studying action potentials is that they appear to be essentially identical, regardless of whether we are recording membrane currents or changes in membrane potential. This is because the excitable ion channels that cause them momentarily dominate the membrane conductance and, therefore, the changes in membrane potential observed effectively mirror fluctuations in current.

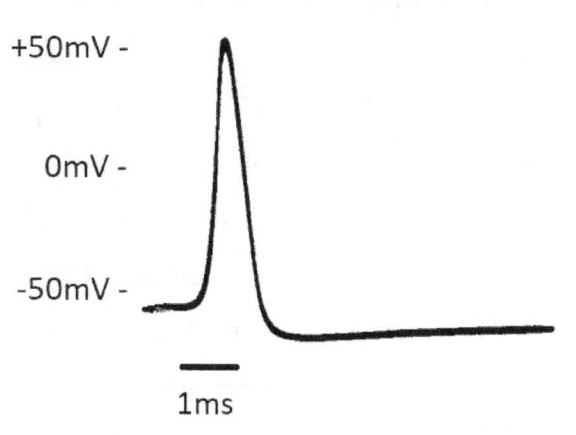

Inset is a representation of an early recording of an action potential obtained from a squid giant axon.[28] Note, even at low temperatures, that action potentials are (a) extremely fast, arising within a millisecond; (b) inherently regenerative, essentially opening, closing and 'resetting'; and (c), that the membrane potential of the axon is initially driven towards the equilibrium potential for Na+ (E_{Na}) and then, within a thousandth of a second, dramatically 'rebounds' towards the equilibrium potential for K+ (E_K).

The early pioneers of electrophysiology did not take long to deduce that such a phenomenon required the explosive entry of positive charge into the axon immediately followed by an explosive efflux of positive charge from the axon. Ionic substitutions within the sea water bathing the squid giant axon soon led them to the conclusion that the ionic species mediating these rapid changes in membrane potential must be Na+ and K+, respectively. This, in turn, led to the realisation that the Na+ and K+ ion channels underlying these action potentials must be sensitive to changes in membrane potential and have the properties of being able to rapidly activate, inactivate, and regenerate their capacity to fire again.

If we look at the phases of the action potential again more closely, we note that there are four distinct *(albeit fleeting)* phases, each of which marks a very important kinetic event within the evanescent life of an action potential.

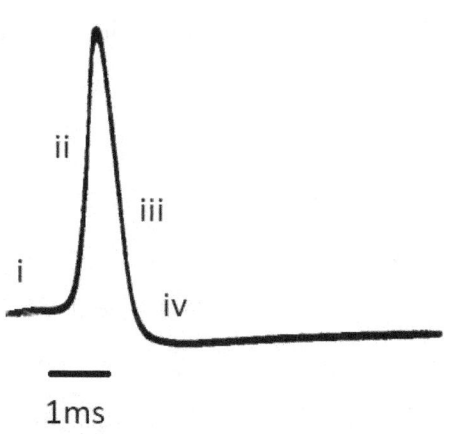

The first, (i) is an initial gradual depolarisation towards the excitable *threshold* for Na+ channel activation, at

---

[28] This is a reproduction rather than an actual recording.

which point all available Na+ channels are recruited in concert. This evokes a runaway depolarisation of the membrane potential (ii) which, in turn, triggers two further rapid events, namely the rapid inactivation of the population of excitable Na+ channels and the rapid activation of a population of excitable K+ channels which then serve to elicit a swift repolarisation of membrane potential (iii), which then momentarily overshoots the original resting membrane potential (iv), leading to what is known as the 'after-hyperpolarisation'. During this phase, the excitable membrane of the axon is said to be *'refractory'* and no further action potentials may be triggered. This refractory hyperpolarisation enables the Na+ channels to be 'reset' and the Na+-K+ ATPase to actively re-establish the asymmetrical gradients of K+ and Na+ which make action potentials an electrochemical possibility.

How fast are action potentials?

The fastest recorded action potentials, in terms of their duration, occur within a millisecond, although their kinetics can be slowed by neuromodulators *(or nervous exhaustion)*. With respect to their frequency, thick-tufted neurons within the primary somatosensory cortex have been recorded firing at frequencies of over 100Hz; whereas the dopaminergic pacemaker neurons of the ventral tegmental area (VTA) which I recorded from at the University of Durham, may *spontaneously* 'fire' at frequencies as low as 1-8 Hz, increasing to over 20Hz upon stimulation - although many neurons only fire when they receive and sum excitatory inputs.

In terms of their *conduction velocity* (*i.e.* speed of transmission along the axon), the fastest recorded axon potentials arise within spinal motor neurons and can travel at speeds of up to 120 m/s, whereas the slowest known action potentials are found within nociceptors and have a somewhat more pedestrian rate of passage of around 0.5 m/s.

The information carried by action potentials may be encoded within the form of their wavefront (*i.e.* the shape and duration of the depolarising wave which serves to modify synaptic integration and transmission), their frequency, and their firing pattern (many action potential trains arise in bursts). The topic of neural encoding is however too vast and complex to be considered here, although neurons may be considered, at a simplistic level, to be biological *transistors*, comparing one input (expected) to another (observed) and reporting *any differences* arising through a common integrated output.

How do we record action potentials?

Action potentials are such large electrical events that they can be recorded by a variety of methods, including; intracellularly using sharp electrodes (or even wires in the case of the squid giant axon); extracellularly using multi-electrode arrays or sharp electrodes 'in the bridge mode' of the amplifier; and, of course, using patch clamp electrodes in either the current clamp or voltage clamp mode *(although this requires specialized considerations as detailed below)*.

There are significant limitations and advantages to each recording method. For instance, although extracellular electrodes placed near to the surface of the membrane only record action potentials as events, they do not interfere with the underlying physiological processes. We will not, however, have time to consider them all in detail here.

How can we distinguish between excitable Na+ and Ca2+ channels?

It should be noted that action potentials are not exclusively mediated via excitable Na+ and K+ channels, as regenerative currents can also be carried via Ca2+ ions. A great variety of Ca2+ channels have evolved to serve a wide range of specialized functions, coupling changes in membrane excitability to a number of cellular events including muscular contraction, cellular movement, exocytosis, synaptic transmission, and neuronal plasticity (memory). As we shall discover in the next chapter, many more potentially remain to be discovered.

Due to their rapid kinetics, it is sometimes difficult to distinguish action potentials conducted by the entry of Ca2+ ions into the cells from those mediated via the influx of Na+ ions. Fortunately, or rather unfortunately, nature has provided us with a formidable array of neurotoxins which target specific types of Ca2+ channel. Given the central importance of excitable Na+ and Ca2+ channels, the existence of highly specific toxins which act upon one specific class of channel has proven immensely useful in the characterisation of their role in cellular excitability.

The most famous and, to date, perhaps the most useful of these has been tetrodotoxin (TTX), a highly selective inhibitor of excitable Na+ channels. However, as we shall see, no ion channel inhibitor, not even a toxin isolated from a powerful venom, may be perfectly selective.

There are of course less specific inhibitors of K+ and Ca2+ channels which have been synthesized, although these tend to act by blocking the selectivity filter at the mouth of the channel and are therefore (as we shall see), prone to be rather more promiscuous in their actions. One of the more successful classes of Ca2+ channel - and least toxic - have been the dihydropyridines (DHPs), which tend to modulate the general excitability of neurons by blocking L-type Ca2+ channels. However, this does not always make them good clinical therapeutics. For instance, in my experiments, while dopamine reversibly depressed the firing rate of VTA neurons by acting on autoreceptors, the addition of the DHP nifedipine and dopamine caused an *irreversible cessation* of the spontaneous firing of VTA neurons. However, it is difficult to envisage a clinical application for permanent *anhedonia*.

Do dendrites contain excitable ion channels?

As we continue to explore the intricacies of the central nervous system, the closer we look, the more complex neuronal processes seem to become. One such revelation was the discovery that excitable Na+ and Ca2+ channels are also present within the fine dendritic processes that form the receptive field of neurons. This fine dendritic arbour provides the cell body of the neuron, or *soma*, with its sensory inputs. The presence of excitable ion channels within the dendrites indicates, at least in some classes of neuron, that information is *actively* rather than *passively* propagated from synapses to the soma.[29]

This revelation has profound implications for computation neurobiology, as synaptic inputs which are propagated purely electrotonically *(that is to say passively without being carried by action potentials)* along the fine dendritic processes would be expected to attenuate rapidly with distance, and thus our understanding of the nature of the summation and integration of synaptic

---

[29] Electrical and Calcium Signaling in dendrites of Hippocampal Pyramidal Neurons. Jeffrey Magee, Dax Hoffman, Costa Colbert, Daniel Johnston. Annual Review of Physiology 1998 60:1, 327-346

inputs has changed profoundly, given that synaptic inputs are conducted actively in the form of regenerative action potentials from dendritic synapses to the soma.

The challenges of recording currents from neurons

Other than the technical issues of isolating intact neurons or maintaining them in culture, neurons can prove to be very challenging to record from for several reasons.

The first is that they are often very large, which means that their input (membrane) resistance is often very low. In some neurons, the input resistance[30] can be as low as 50MΩ.[31] This means that voltage clamp recordings will be severely distorted if the series resistance (Rs) is significantly greater than 1 or 2MΩ, which means that we effectively need to use 'giant' patch electrodes which makes the practical reality of obtaining a high resistance seal an issue. If the input resistance of a neuron was 50MΩ, then a typical patch electrode, with a resistance of 5-8MΩ, would instantly create a series resistance error of over 10% which, when added to the fact that the dendrites and axon effectively form distinct electrical compartments which are beyond the 'command' voltage of the electrode (*i.e. they are not voltage clamped*), means that measurements of current are likely to be both severely distorted and attenuated (see also Appendix C).

It has been argued, when recording from neurons with short dendrites, that the absence of a perfect space clamp does not significantly distort the properties of the action potentials or currents measured. This contention, in conjunction with passive cable theory, has been employed in an attempt to lend 'validity' to voltage clamp experiments performed with neurons that possess extensively branched dendritic arbours, wherein authors have contended that the decay in potential arising is sufficiently shallow to permit an effective space clamp of the neuron.

However, others have shown that voltage-gated K+ and Ca2+ currents are significantly distorted, even within neurons with relatively short dendrites. This arises due to the 'shunting' of the

---

[30] Note that this is not a constant.
[31] *e.g.* Lee et al., A disinhibitory circuit mediates motor integration in the somatosensory cortex. Nature Neuroscience 16, 1662–1670 (2013).

effective reversal potential during the activation of excitable channels. Thus, recordings of voltage-gated currents recorded from branching neurons are invariably and substantially distorted.[32]

As the dendritic surface area of a neuron may be a hundred-fold greater than that of the soma, when we attempt to apply a voltage clamp via the soma, the dendritic 'compartment' is neither space clamped, nor voltage clamped. However, in the following experiments, this constraint effectively plays to our advantage. When we record from the soma using a single electrode, we are able to infer that excitable channels, capable of supporting action potentials, must be present within the dendrites if multiple action potentials are found superimposed within the recording.

If we voltage clamp the soma, we can reasonably suggest that only one action potential can be elicited at any one point in time within the soma and residual length of the axon within a slice preparation. Thus, if we record from retinal ganglion cells (which typically have an input resistance of approximately 125MΩ) using electrodes with an access resistance of 1-2 MΩ, then we largely eliminate issues arising from the series resistance error.

Another limitation to consider is the sampling frequency which, given that action potentials occur on a millisecond time scale, means that we need to be recording with a sampling frequency greater than 4kHz.

It has become clear that most neurons are not *isopotential*, in that different regions of the axon, soma and dendritic arbour may experience very different levels of ion channel activity and membrane potential at any given instant. This property has been extensively studied using low resistance patch clamp electrodes to minimise the attenuation of the dendritic signal through space clamping artifacts. When the soma is voltage clamped and depolarized by brief command potentials, the regenerative current spike caused by the opening of voltage-gated sodium channels can be observed to separate into distinct population peaks. Regehr and Armstrong[33] duly inferred that, because threshold potential in the soma/axon hillock would result in an all-or-none spike of

---

[32] Dan Bar-Yehuda & Alon Korngreen. Space-Clamp Problems When Voltage Clamping Neurons Expressing Voltage-Gated Conductances. Journal of Neurophysiology Published 1 March 2008 Vol. 99 no. 3, 1127-1136.
[33] Regehr W, Kehoe JS, Ascher P, Armstrong C. Synaptically triggered action potentials in dendrites. Neuron. 1993 Jul;11(1):145-51.

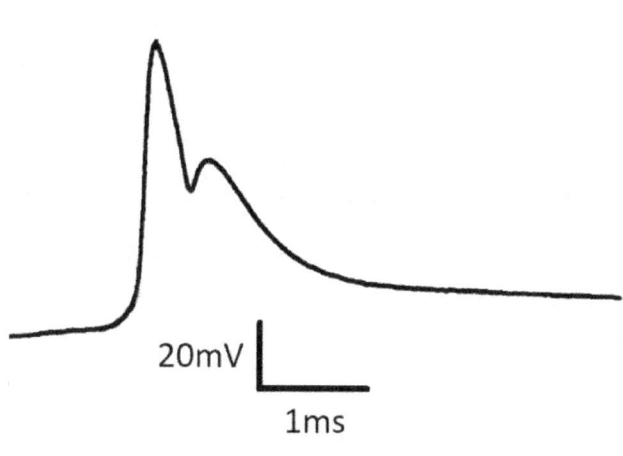

excitable sodium current if excitable sodium channels were present exclusively in the axon and soma, then excitable sodium channels must therefore also be present in the dendrites, supporting the earlier premise of Sugimori and Llinas.[34]

Now we come to the question at hand. Could I demonstrate that the dendrites of retinal ganglion cells also contain regenerative Na+ and/or Ca2+ channels by recording from the soma?

To test this hypothesis, whole cell voltage clamp recordings were obtained from 150μm sections of the Tiger salamander retina (*Ambystoma tigrinum*) and subsequently from a flat-mounted retinal preparation to prevent damage to the dendritic layer caused by sectioning. Voltage clamp recordings were made at the level of the soma using 1-2MΩ electrodes and the cells were routinely voltage-clamped at -50mV. Brief depolarising voltage steps (3-4mV, 100ms) were applied in increments using P-clamp software and acquired at the limit of resolution of the board (4kHz). The threshold for the main regenerative spike was reached at around -33mV, but before this potential was reached, smaller regenerative inward currents were elicited with a latency that became shorter latency as the somatic threshold potential was approached *(not shown)*. As this somatic threshold was passed, the latency of the smaller regenerative current decreased so that the two peaks that were separated at or near threshold, eventually merging into one as the magnitude of the depolarizing pulse became large enough to activate all the excitable channels simultaneously *(not shown)*. This smaller spike most likely represents a population of excitable channels which are present within the dendrites with a higher (more negative) threshold potential.

A current clamp recording obtained from the soma of a retinal ganglion cell is shown *inset*, in which more than one action potential may be seen arising in response to the injection of depolarising current into the soma. We might reasonably infer that the second smaller, later and

---

[34] R Llinás and M Sugimori. Electrophysiological properties of in vitro Purkinje cell dendrites in mammalian cerebellar slices. J Physiol. 1980 Aug; 305: 197–213.

slower action potential has arisen within the dendritic field. In this recording, we see only one additional action potential within the recording although, in others, evidence of multiple action potentials was observed *(not shown)*.

Thus, we can conclude that, as would be reasonably predicted, that retinal ganglion cells are *anisopotential* and that their dendritic field contains excitable Na+ and/or Ca2+ channels.

Chapter Seven

The kinetics of action potentials

In the previous chapter we sought to introduce the fundamental properties and nature of action potentials. In this chapter, the experiments delve more deeply into the kinetic behaviours of the underlying excitable channels.

In these experiments, I used a combination of ionic gradients, voltage protocols, and pharmacology to attempt to study the properties of the excitable ion channels which are present within the pancreatic β-cell. However, before we address the complexities of kinetic analysis, we must first consider the biological question we are attempting to answer.

The specialised endocrine cells of the pancreatic islets and adrenal gland share many features in common with the synaptic terminals of neurons, although they release hormones rather than neurotransmitters in response to excitation. As in nerve terminals, excitatory responses are mediated by the activation of voltage-sensitive Ca^{2+} channels which leads to the entry of Ca^{2+} into the cell and the fusion of vesicles containing the signal (*i.e.* a neurotransmitter or a hormone) with the membrane to enable the release of the signal.

However, most previous recordings made from excited pancreatic β-cells appeared to show slow waveforms rather than the rapid action potentials which are characteristic of recordings obtained from intact islets *(which are also electrically coupled by gap junctions)*.[35] As is discussed in the preceding chapter, we are accustomed to seeing rapid openings and closures of excitable calcium channels which, in the case of the pancreatic β-cell, trigger the release of insulin in response to elevations in blood glucose.

---

[35] Loppini A, Braun M, Filippi S, Pedersen MG. Mathematical modeling of gap junction coupling and electrical activity in human β-cells. Phys Biol. 2015 Sep 25;12(6):066002.

This anomaly understandably led to some debate about the properties of Ca2+ channels present within the pancreatic β-cell membrane, and so my experiments sought to shed light on their underlying nature, regulation and kinetic properties.

The prevailing consensus was that a population of voltage-dependent L-type Ca2+ channels played a pivotal in stimulus-secretion coupling within the pancreatic β-cell. Given that small clusters of pancreatic β-cells have served as a favoured model for the electrophysiological study of pancreatic β-cells, this preparation was selected as our primary experimental model[36] for a further investigation into why fundamental discrepancies have arisen between observations of the kinetic behaviour of currents elicited from β-cells in response to glucose in intact pancreatic islets and cultured single β-cells. The addition of D-glucose elicits regenerative action potentials in islets at a steady state plateau potential, yet currents routinely recorded from single isolated β-cells were slow and transient in nature.[37] [38] [39] So what becomes of these regenerative currents[40] when we patch clamp β-cells?

There was a second phenomenon which I was asked to address. If L-type Ca2+ channels burst continuously in response to glucose, then why is insulin secretion pulsatile rather than continuous in nature? Does another class of Ca2+ channel exist within the membrane of the β-cell membrane with properties that could account for the pulsatile nature of insulin secretion *in vivo*?

Small clusters of β-cells were adopted as our model system, serving as an intermediary between single cells and intact islets which are electrically coupled and therefore too large to voltage clamp. The perforated patch clamp technique was again used to avoid the dialysis of key

---

[36] As they minimise the complications of global oscillatory changes in input and gap junctional resistance within islets.

[37] Barnett, D.W., Pressel, D.M., Misler, S. (1995). Voltage-dependent Na$^+$ and Ca^{2+} currents in human pancreatic islet b-cells: evidence for roles in the generation of action potentials and insulin secretion. Pflugers Arch-Eur.J.Physiol. 431: 272-282.

[38] Smith, P.A., Rorsman, P., Ashcroft, F.M. (1989). Modulation of dihydropyridine-sensitive Ca^{2+} channels by glucose metabolism in mouse pancreatic b-cells. Nature. 342: 550-553.

[39] Bokvist, K., Eliasson, L., Ammala, C., Renstrom, E., Rorsman, P. (1995) Co-localization of L-type Ca^{2+} channels and insulin-containing secretory granules and its significance for the initiation of exocytosis in mouse pancreatic b-cells. EMBO J., 14: 50-57.

[40] Note that we are careful to call them regenerative currents rather than action potentials as, strictly speaking, these are currents recorded at a fixed holding potential rather than changes in membrane potential.

cytosolic components and again I employed intricate voltage step protocols, ion substitutions, and inhibitors to isolate and characterize the Ca2+ channels present within pancreatic β-cells.

Using incremental voltage steps to separate voltage-gated ion channels

A simplified voltage step protocol was applied to small clusters of islet cells held at a resting potential of –70mV (termed the 'IK1' protocol). After a lengthy 40-60 min pre-incubation in Krebs media containing 3mM D-glucose (G3), only outward currents could be observed using *pseudo-physiological* ionic gradients, *as shown below*.

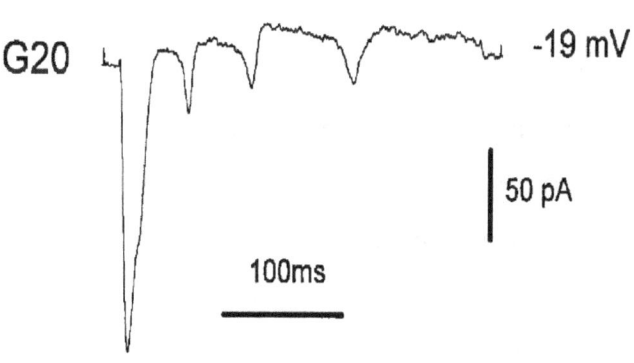

The upper recording shows a typical current response elicited by a 300ms depolarizing step to -28mV from a holding potential of -70mV after pre-incubation in 3mM D-glucose (G3) and the lower recording shows a voltage step to -19mV following a 2-minute incubation in 20mM D-glucose (G20). The IK1 protocol consisted of a series of 300ms depolarizing steps in 3mV increments from a holding potential of –70mV with a 2s interpulse interval.

Thus, as in intact islets, small β-cell clusters respond to depolarisations of the membrane with a train of *regenerative inward* current deflections which occur in parallel with a suppression of the outward current, which is known to be carried by ATP-sensitive K+ (K_{ATP}) channels.

The threshold for voltage activation of the train of recurrent deflections was dependent upon the concentration of D-glucose, with a mean threshold of -47.2 ± 1.5mV (n=30) after a 2-minute incubation in G20, and -26.5 (± 1.5) mV after 2 min incubation in G5 (n=4, $P<0.005$ ANOVA).

These recurrent deflections were of lower threshold than a ***second, larger transient*** voltage-activated deflection (threshold -35.5 ± 2.3mV (n=6) in G20 after 5 minutes of incubation).

Glucose-activated current deflections exhibit strongly voltage-dependent kinetics

Here we address four key parameters of excitable channels, namely their frequency, open probability, activation threshold and latency. As previously discussed, the *frequency* of action potentials is one of the principal ways in which information is encoded and transmitted within the nervous system, and is typically expressed in cycles per second, or *Hertz*. The amplitude of the current spike is again dependent upon the product of unitary conductance, the number of ion channels present, and their open probability. The *open probability* of excitable ion channels is in turn dependent upon the *kinetics* of their activation and inactivation. The *threshold* is that potential at which the voltage sensor present within excitable ion channels gates, or opens the ion channel. The *latency* is the delay between the onset of the voltage step and channel activation.

The threshold, frequency, amplitude, and latency of the smaller glucose-activated recurrent inward deflections (*i.e.* those which fire in a continuous train without evidence of a decrease or *accommodation* in their rate), which were typically of 10-15ms in duration, appeared to be strongly influenced by the holding potential. This is to say that their activation kinetics in the presence of glucose are strongly voltage-dependent.

Unlike the excitable Na+ channels that are present throughout the nervous system, these recurrent inward deflections show no apparent requirement for membrane repolarisation for their reactivation. The absence of any evidence of distant 'unclamped' current deflections, or 'echoes' *(which might be expected if unclamped neighbouring cells were being depolarised to threshold by the metabolic action of D-glucose)* is consistent with the argument that an effective spatial voltage clamp, or space clamp, had been obtained in these recordings *(this will be discussed in greater detail later in the chapter).*

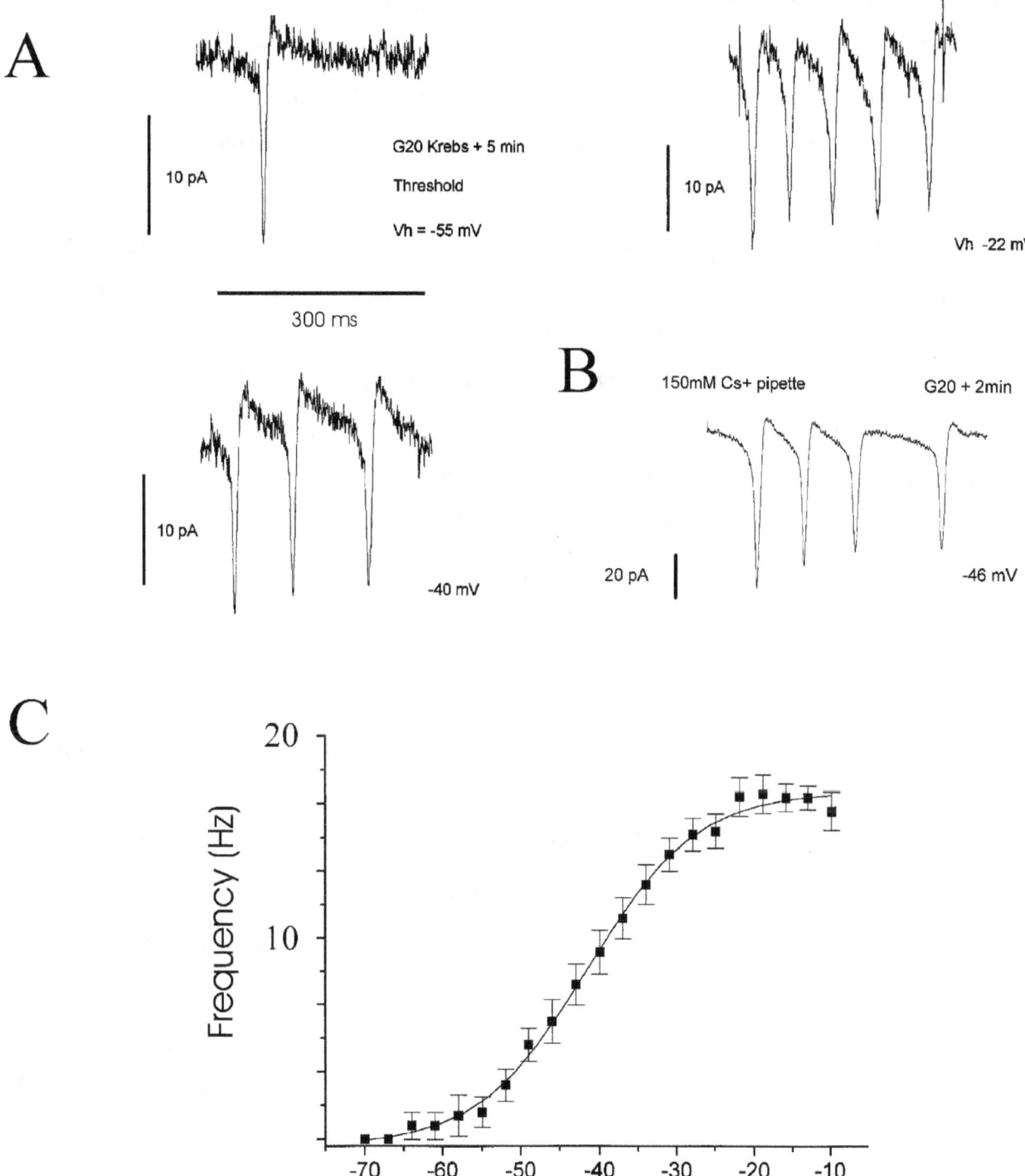

The figure presented above below shows how the frequency of these regenerative inward current deflections increases as a function of holding potential.

The recordings presented within the upper panels illustrate how the frequency of the inward current deflections activated by 20mM D-glucose (G20) appear to be strongly voltage-dependent. These are currents (panel A) that were elicited by a series of incremental 300ms voltage steps from a holding potential of -70mV to -55mV (the activation threshold for this particular recording), and subsequent steps to -40mV and -22mV within the same IK1 protocol. When the pipette solution was changed from K+ to Cs^+ (panel B), which is a permeant inhibitor [41] of K+ channels, the appearance of the regenerative currents was apparently unaffected (B).[42]

Note that the frequency of the regenerative inward current deflections evoked by G20 is steeply voltage-dependent (C), with their sensitivity to voltage being at its 'steepest', or most acute, between the holding potentials of -50 and -20mV, with a half-maximal frequency ($V_{0.5}$) elicited at -43mV. The mean frequency of inward current deflections was plotted as a function of the step potential (Vstep) for regenerative inward current deflections (means ± SE from 7 separate β-cell clusters) with a fit to the Boltzmann equation *(which describes the thermodynamic behaviour of a system which is not in equilibrium)* in the following form:

$$P_0 = 1/(1 + \exp\{q(V0.5 - V)\beta\})$$

Where Po is the probability and V is the holding potential. Evidently, the rate of activation of the intrinsic voltage sensor within the channel complex is strongly voltage-dependent and can be described by a fit to the Boltzmann equation.

Note that the latency between the *onset of depolarization* (i.e. the time at which the voltage step is applied), and the appearance of the first, second and third recurrent current deflections also appears to be voltage-dependent.

The latency can in effect be used as a measurement of the sensitivity, or rate of activation of the channel's intrinsic voltage sensor. The fits to the 1st, 2nd and 3rd latencies of the 1st, 2nd, and 3rd

---

[41] Cs+ is a larger monovalent cation from the same family of the periodic table as K+. Thus, it enters the channel pore but effectively occludes the channel pore due to its slower permeation, if indeed this occurs.

[42] There is no evidence to suggest that the inward current deflections incorporate a substantial K^+ component that masks or distorts their true kinetics, as (i) there was little current 'overshoot' from baseline following the inactivation phase of the current deflection; (ii) the current deflections were fast inactivating, minimizing attenuation of current amplitude through any co-localized Ca^{2+}-activated K^+ channels, and (iii) the kinetics of the spikes were not substantially affected by intracellular Cs^+ (as shown above), an inhibitor of voltage-dependent K^+ currents.

spikes also indicates that the inter-deflection interval, or period between spikes, decreases as a function of voltage. This is plotted as a function of step potential in the graph below.

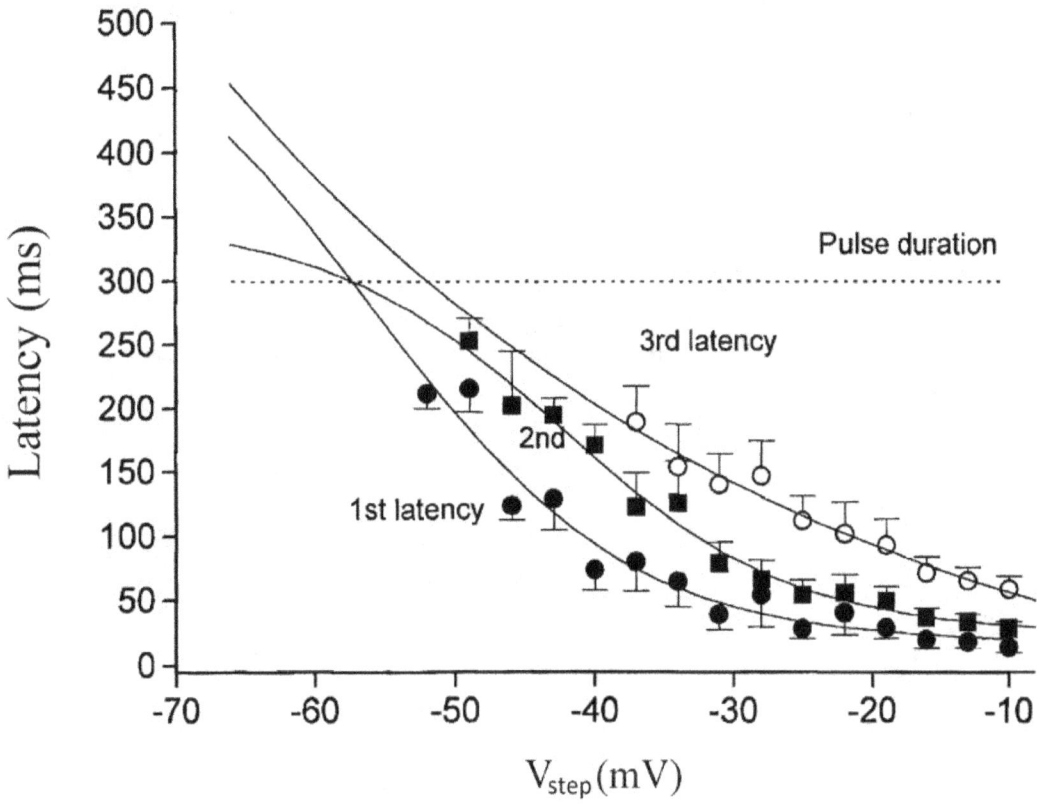

Values in the graph presented above are means ± S.E.M. from 15 clusters with 150mM KCl in the recording electrode and all fits are to the Boltzmann equation.

Establishing the ionic selectivity of the inward deflections

The ionic selectivity of these glucose-activated recurrent inward current deflections should be readily apparent from shifts in the amplitude of these current deflections after ionic substitutions to change the equilibrium potentials for Na+ and Ca2+.

The next figure shows representative recordings from a small β-cell cluster in which the external Na^+ was reduced from 144 to 59 mM by the equimolar substitution of NaCl with NMDGCl (n=4) or [Ca2+]o was reduced from 2.5 to 0.25mM by the equimolar replacement of

$CaCl_2$ with $MgCl_2$ (n=6). Clearly, the amplitudes of these recurrent deflections are reversibly diminished by Ca^{2+} substitution but not Na+.

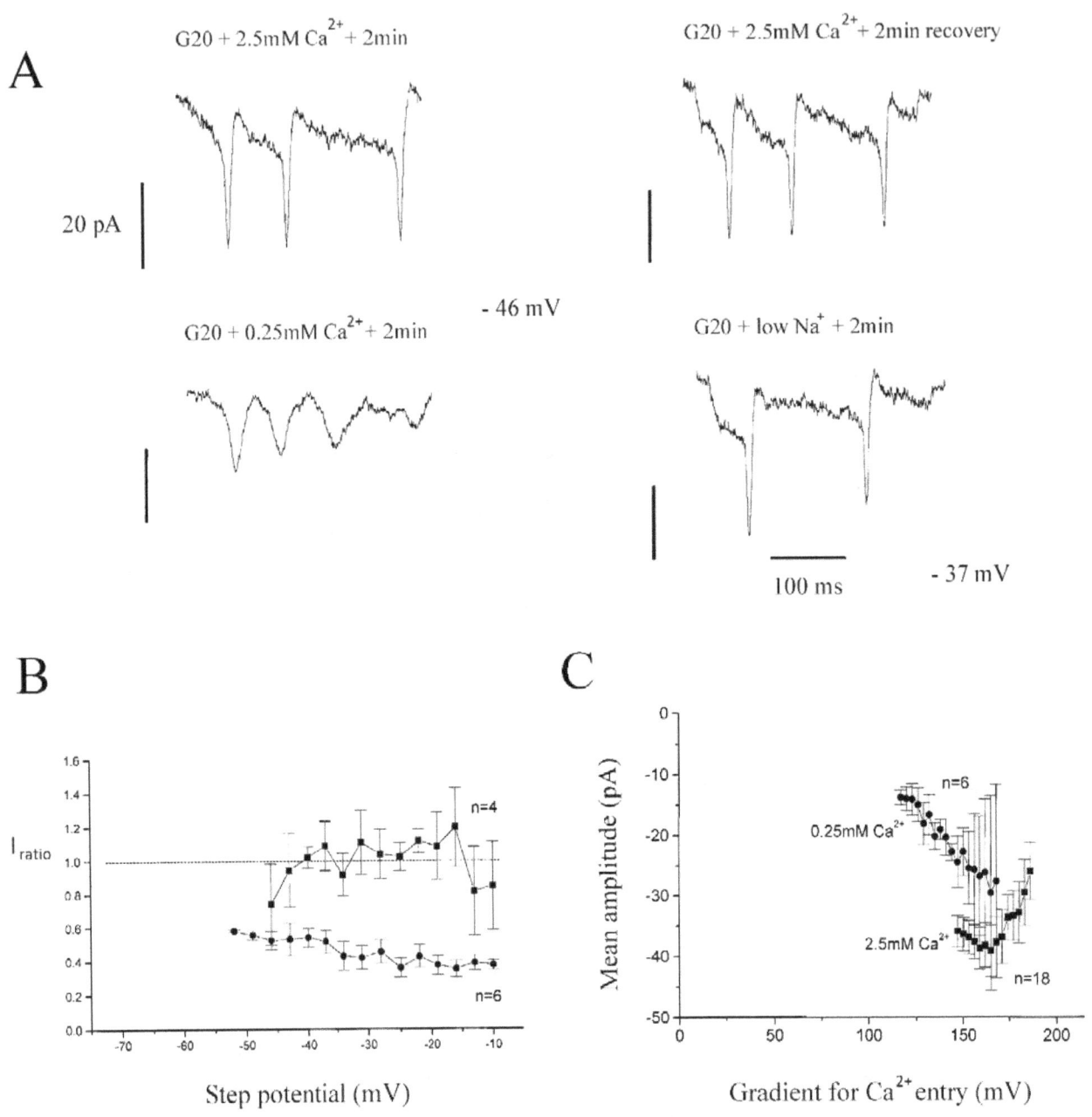

The figure shows the current responses elicited from a small voltage-clamped β-cell cluster bathed in Krebs medium containing 144mM Na+ and 2.5mM Ca2+ (A) in response to a 300ms depolarizing step to -46mV from a holding potential of -70mV after pre-incubation in G20 for 2 min when the cluster was bathed first in 2.5mM Ca2+ *(upper trace)* and then 0.25mM Ca2+ *(lower trace)*. The absence of any change in regenerative current amplitude (or duration) upon equimolar replacement of extracellular Na+

from 144 to 59mM with the impermeant cation NMDG⁺ is also shown in the right-hand pair of recordings. The pipette solution used was Cs150. The ratio of the peak current deflections observed upon ionic substitution at each step potential are shown for both Na+ and Ca2+ in panel B. If we then plot the peak conductance of regenerative current deflections as a function of the calculated electrochemical gradient for Ca2+ entry for both 2.5mM Ca2+ and 0.25mM Ca2+ (assuming an intracellular free Ca^{2+} activity of 200nM) we observe that the amplitude of the inward currents observed is determined by the electrochemical gradient for Ca2+.

These ion substitution experiments clearly indicate that Ca2+ is the dominant permeant cation through this conductance pathway in the presence of *pseudo-physiological* ion concentrations. However, we should bear in mind that the amplitudes of the peak current deflection (I) is given by;

$$I = n.Po.i$$

where *n* is the number of functional channels, *i* the unitary conductance and P_o is the mean open probability *(although again none of these parameters can be assumed to be constant)*.

The demonstration that a ten-fold reduction in external calcium to 0.25mM reduces the deflection amplitude by only 40% is perhaps not surprising, as the estimated E_{Ca}, as calculated according to the Nernst relation only changes from +137 to +107mV (see also Appendix F). Equilibrium potentials for Ca^{2+} were calculated according to the Nernst equation in the form:

$$ECa = \frac{RT}{2F}.\ln\left(\frac{[Ca]o}{[Ca]i}\right)$$

Where E_{Ca} is the equilibrium potential for Ca, R is the general gas constant (8.315 J.K⁻¹mol⁻¹), T is the absolute temperature (K), F is Faraday's constant (9.648 x 10⁴ C.mol⁻¹), [Ca]o is the extracellular Ca2+ concentration and [Ca]i is the intracellular Ca^{2+} activity.

Effects of Ca2+ channel inhibitors and chelators upon the recurrent deflections

The inward current deflections observed indicate that there are at least two types of voltage-dependent current activated by glucose. Both forms of inward current deflections were inhibited by the Ca2+ channel antagonist D-600 (100µM) in a voltage-dependent and reversible manner (n=4, *not shown*). The voltage-gated Na$^+$ channel inhibitor tetrodotoxin (TTX) at 100nM had no effect on the frequency, amplitude or threshold of the recurrent regenerative currents, although these regenerative current deflections were abolished following a 60-minute preincubation with the Ca2+ chelators BAPTA-AM (5-50µM) or Fura-II-AM (1µM). These observations suggest that this recurrent Ca2+ channel activity is acutely sensitive to the degree of intracellular calcium buffering, and this is further evidenced by a reduction in their frequency and an increase in their duration when the extracellular Ca2+ concentration is reduced from 2.5 to 0.25mM *(see recordings above)*.

Using pharmacology to distinguish between ion channel populations

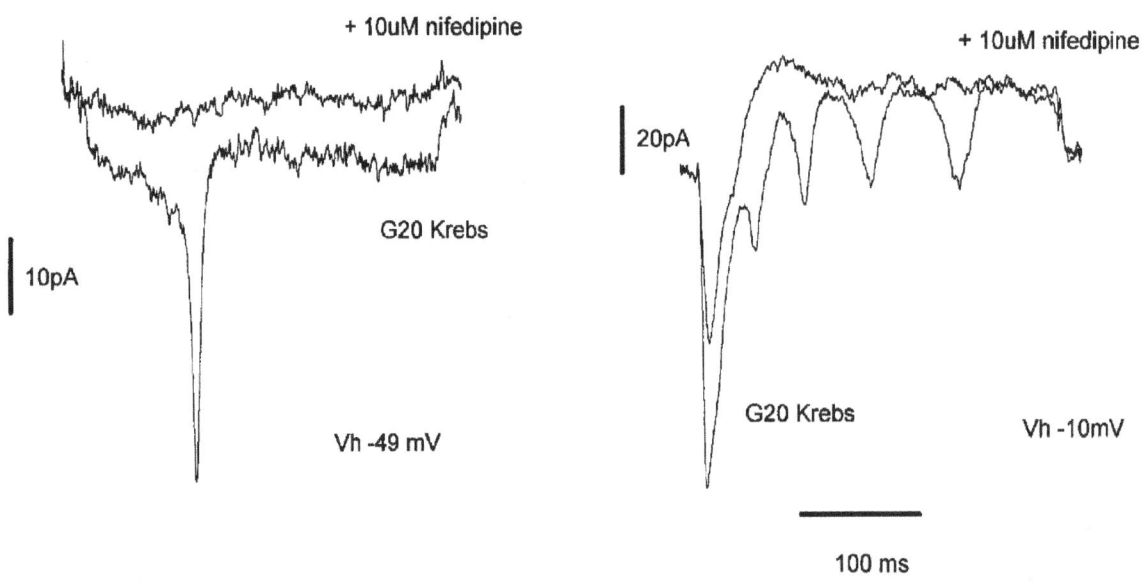

Pharmacology reveals two distinct glucose-activated Ca2+ currents. The recordings inset above show that the L-type Ca2+ channel nifedipine inhibits the recurrent, but not the high threshold, glucose-activated current deflection in small β-cell clusters. Paired recordings of current deflections elicited are shown after a 2-minute pre-incubation in G20 using the IK1 protocol to threshold (-49mV, left-hand trace pairing) and subsequently to -10mV (right-hand trace pairing), both before (lower trace) and after (upper trace) the addition of 10μM nifedipine to the bathing medium (n=6). The pipette solution used was Cs150.

In six separate experiments, the recurrent inward deflections were reversibly abolished in the presence of 10μM, but not 1μM, nifedipine[43], although the *larger, transient* inward current deflection persisted in the presence of this L-type Ca2+ channel inhibitor. By virtue of their kinetic resemblance to the L-type spike trains evoked by D-glucose in whole islets, and their sensitivity to dihydropyridines, we shall refer to these recurrent Ca^{2+} current deflections as being mediated via glucose-activated L-type Ca^{2+} (L_G-type) channels.

Note that the use of nifedipine has enabled us to effectively isolate the higher threshold current deflection that is activated in the presence of G20 at more depolarised potentials. From a kinetic standpoint, this second current could be readily discriminated from the L_G-type channel population by its shorter latency, transient activation, and greater amplitude. This deflection could also be isolated after a 2-minute pre-incubation with 10μM phentolamine (in 4 experiments, see page 81).

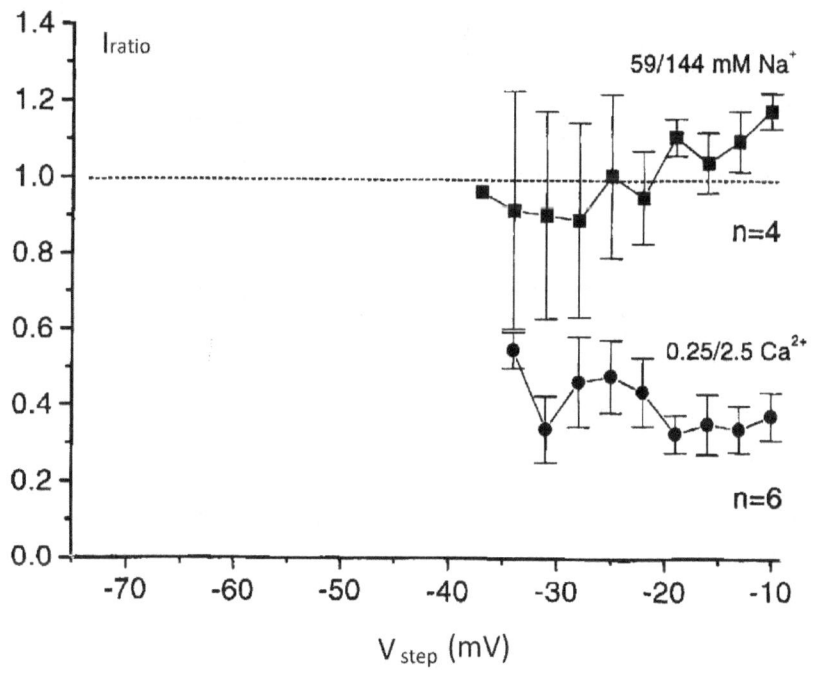

If we then repeat the ion substitution experiments for this larger transient inward current, we note that the ratios of peak amplitudes before and after equimolar the replacement of Na+ remain unaffected *(inset left)*, while reducing external Ca^{2+} to 0.25mM from 2.5mM again reversibly depressed the peak current amplitude in six separate experiments.[44]

---

[43] another member of the dihydropyridine family of L-type Ca2+ channel inhibitors.

[44] Ratios were calculated by dividing the isopotential peak deflection amplitude in the reduced ion environment by the paired amplitude in the control cationic environment.

Kinetic properties of the high threshold inward current

As may be observed from the graph below, this larger Ca2+-activated current deflection exhibits a sigmoidal relationship between peak current and holding potential. The mean peak current amplitude is presented as a function of step potential from 6 recordings and the fit is to the Boltzmann equation.

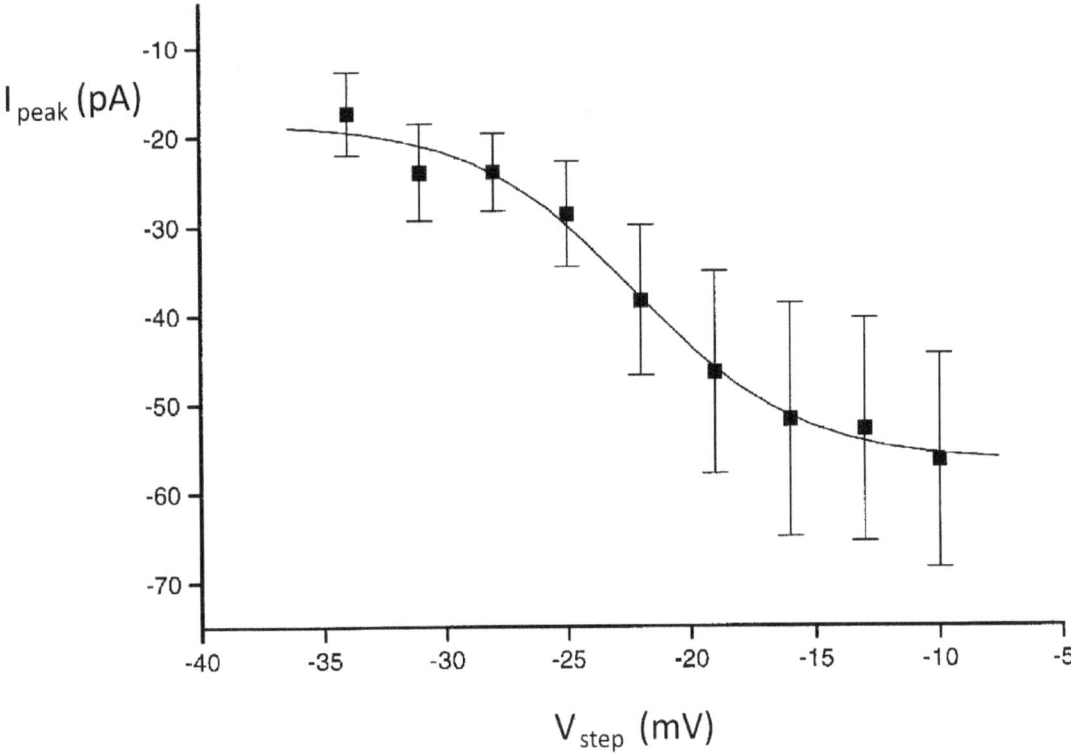

This higher high threshold, nifedipine-resistant current exhibited no evidence of 'run-down' in its amplitude over 10 minutes of recording in G20, and its amplitude could be fully restored or reactivated, upon membrane repolarisation to -70mV following the initial depolarising step. I named this novel Ca2+ conductance the 'G-type' Ca2+ channel, given its unique attributes as a voltage- and glucose-activated and dihydropyridine-resistant Ca2+ current.

The threshold for activation of the G-type current was significantly lower following a 5-minute pre-incubation in G20 (-35.5 ± 2.3mV, n=6 in nifedipine) than after 2 minutes of incubation in

G20 (-26.5 ± 5mV), which is suggestive of a slower rate of activation than is observed for the L_G-type channels. The G-type current exhibited a maximal slope conductance of 2.6nS between -25 and -20mV and a peak amplitude at -10mV.[45]

The recordings also suggested a strong voltage-dependent rectification of the G-type current, with the macroscopic current *increasing* with a *decreasing* electrochemical gradient for Ca^{2+} entry. However, despite its evident Ca2+ selectivity, the G-type current was partially inhibited by 100nM TTX in a voltage-dependent manner between -19 and -25mV and it was completely inhibited in the presence of 1μM TTX (*not shown*). These observations raise the intriguing possibility that the G-type Ca2+ channel may share a common ancestor with the voltage-gated Na+ channel family.

Voltage-dependent properties of the G-type current

The reactivation of the inactivated G-type current requires a repolarization step, which is again reminiscent of the classical kinetic properties of excitable Na+ channels. Thus, the G-type Ca2+ channel would be expected to contain a voltage sensor which gates, or opens the channel. We can effectively study the G-type channel's reactivation kinetics by using *paired* voltage pulse protocols.

This approach is illustrated in the following figure in which the currents elicited by a paired-pulse voltage protocol are used to determine the half-maximal repolarization potential ($V_{0.5}$). Essentially, after applying an initial step depolarisation, the membrane is repolarised in a series of 300ms increments, before a second depolarisation step is applied.

---

[45] thereafter the current declined, but measurement was complicated by the activation of an outward current.

A

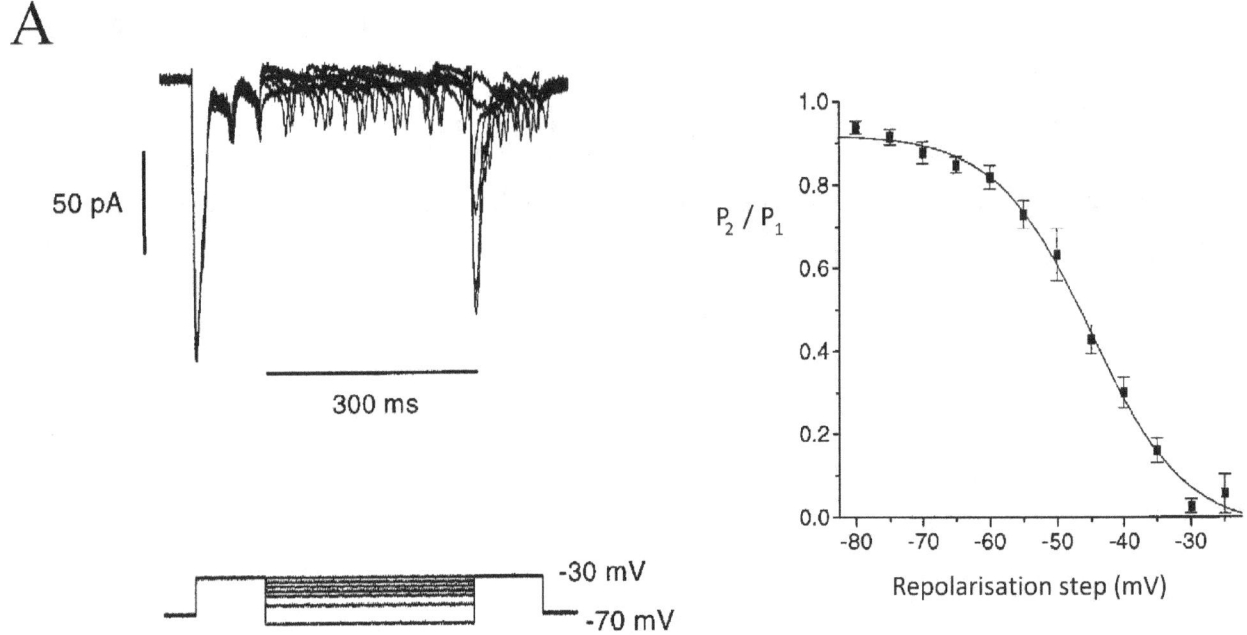

B

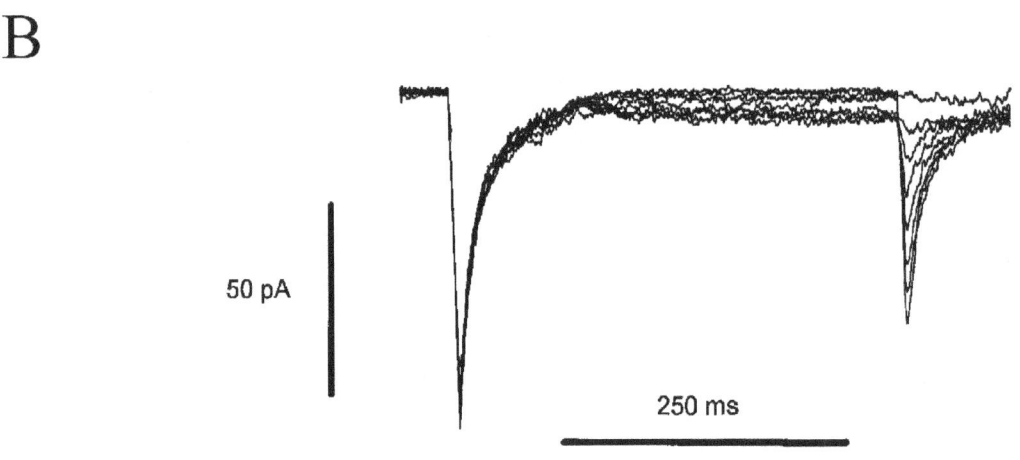

The figure above reveals the voltage-dependent reactivation kinetics of the G-type current. The current traces elicited by a paired-pulse protocol are shown together with the corresponding voltage protocol in panel A. The voltage protocol comprised a 100ms step pulse (V_{step}) to -30mV from a holding potential of -70mV which was followed by an initial 300ms repolarization pulse to -80mV followed by a paired 100ms V_{step} to -30mV. The sequence is then ended by returning to -70mV for a 2-second interval before a subsequent paired-pulse was applied wherein the repolarization step interval was diminished by 5mV for each successive and otherwise identical paired-pulse sequence, thereby giving the G-type current a progressively diminishing repolarization step for each pair of depolarizing pulses applied to the cluster.

The top panel shows seven superimposed current patterns revealing the decay in the amplitude of the 2nd deflection elicited by a paired step depolarization to -30mV concomitant with the progressive reduction in the magnitude of the repolarization step. The corresponding voltage protocol is given below (lower panel). The kinetics of recovery of the G-type current are more clearly observed using the same paired-pulse protocol after a 2-minute preincubation with 10μM phentolamine (panel B).

When we employ a 300ms repolarization step and paired depolarising steps to -30mV, the $V_{0.5}$ for the G-type current in small clusters was -45.0 ± 0.9 mV, with a nearly complete recovery of peak current observed following a 300ms repolarisation to -80mV. The pattern of repolarisation-dependent reactivation is more clearly demonstrated in the presence of 10μM phentolamine, at which concentration the recurrent L_G-type current deflections are inhibited.

The average ratios between the peak G-type currents elicited are presented in the right-hand panel as a function of repolarisation potential.[46] The ratios were calculated by the dividing the peak amplitude of the current elicited by the 2nd 'paired' depolarizing step by the 1st. The data are presented as means (± SEM) from 7 separate experiments. The data points were subsequently fitted to the Boltzmann equation in the form;

$$\frac{P2}{P1} = \left[\frac{\frac{P2}{P1}min - \frac{P2}{P1}max}{1} + e\left(V - \frac{V0.5}{dV}\right) \right]$$

Thus, the extent of reactivation of the G-type current increases with the magnitude of the repolarization step. This would provide a putative 'on-off' switch during the slow wave oscillations which occur in the intact islet in response to intermediate glucose concentrations. We can conclude that the *activation and inactivation kinetics of the G-type Ca2+ channel are consistent with the physiological phenomenon of pulsatile insulin secretion.*

---

[46] Note that the repolarization pulse was of 300ms duration and the pipette solution contained 150mM Cs+.

Voltage-dependence of the glucose-activated Ca2+ channels in isolated β cells

Acutely isolated β-cells exhibited similar current patterns, albeit in a mosaic pattern wherein some single β-cells exhibited only G-type current patterns, whilst others revealed only L_G-type currents, and some exhibited both as presented in the figure below. A kinetic analysis *(not shown)* suggested that the currents is single β-cells were essentially identical to those observed in small clusters.

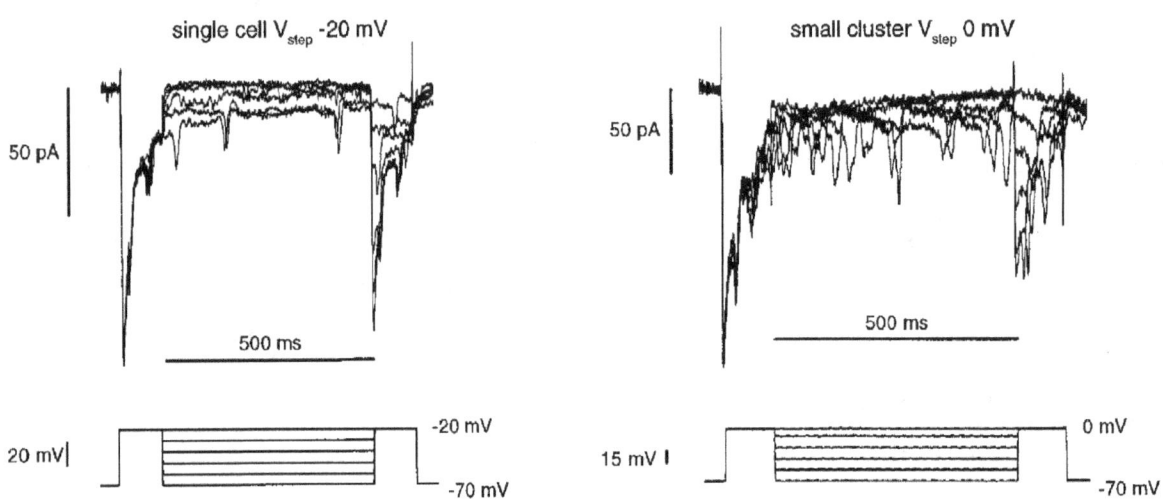

Space clamp considerations

When we record from β-cell clusters the thorny issue of the space clamp resurfaces. The primary advantage of using clusters is that the complications of fluctuations in input resistance and membrane conductance are avoided as small β-clusters appear to 'function better as electrical rather than as biochemical syncytia.' [47]

More importantly, gap junctional conductance has been shown to oscillate synchronously in pairs of acutely dissociated β-cells, although the actual extent of electrical coupling may be greater in intact islets, where *parallel* arrangements of gap junctional conductances would be present. Thus, the coupling coefficients and gap junctional conductances measured between acutely isolated pairs of β-cells at room temperature are almost certainly *underestimates* of their

---

[47] Jonkers, F.C., Jonas, J.C., Gilon, P., Henquin, J-C (1999). Influence of cell number on the characteristics and synchrony of Ca^{2+} oscillations in clusters of mouse pancreatic islet cells. J. Physiol. 520(3): 839-49.

'true' values within small clusters of β-cells, as the rates of cellular metabolism increase non-linearly with temperature, and gap junctional conductance has been shown to increase with the rate of glucose metabolism in β-cells.

Calculating the coupling coefficient

The gap junctional coupling coefficient (CC) arising between two conjoined cells connected by a single gap junctional conductance (g_j) may be represented by the following equivalent circuit *(inset right)*, wherein two parallel RC circuits are formed by the two conjoined cells, each with its own membrane conductance (g_1 & g_2) and membrane capacitance (C_1 & C_2). Note that the membrane conductances and the gap junctional conductance (g_j) are represented by the symbol for a *variable resistor* as they cannot be assumed to be constant.

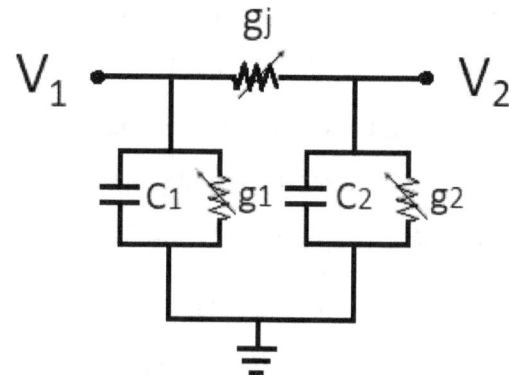

For a given magnitude of *current injection* into cell 1, the gap junctional coupling coefficient CC is given by the relationship;

$$CC = \frac{\Delta V2}{\Delta V1} = \frac{gj}{gj + g2}$$

Where ΔV1 is the change in membrane potential of cell 1 in response to current injection and ΔV2 is the measured change in membrane potential of cell 2 in response to the injection of a known current pulse into its immediate neighbour.

In essence, gap junctions are electrical synapses and operate as low pass filters as the membrane capacitance takes time to charge.

The gap junctional coupling coefficient is thus a central concern to the interpretation of these experiments, as it determines the length constant τ and therefore our ability to spatially voltage-clamp, or 'space clamp', small β-cell clusters. Although 'cluster size' might be approximated to between 3 and 7 cells as judged from isopotential current deflection amplitudes elicited relative to those of identified single cells, in the absence of fluorescent nuclear staining after recordings this cannot be determined directly.

As small clusters do not generally support slow wave oscillations in response to elevated concentrations of D-glucose, they can, therefore, be suggested to remain within the 'active' phase of the response to glucose which is associated with a higher membrane conductance and a greater degree of intercellular coupling. The extensive gap junctional coupling arising between pancreatic β-cells has been shown to be increased by glucose[48] and the intercellular gap junctional conductance oscillates synchronously with membrane potential, with the coupling coefficient (0.74) and gap junctional conductance measured between isolated pairs of β-cells (514 ± 137pS) being considerably higher within the active phase.[49]

However, despite the persistence of an active phase and the extensive parallel gap-junctional coupling conductances (G) arising between clusters containing multiple gap-junctionally coupled cells ($G_T = G_1+G_2+G_3+G_n...$), there are still 'space-clamping' considerations which have been only been addressed empirically here. However, other investigators have addressed this issue by recording from 'displaced' β-cells found at the surface of intact islets which has led to an estimated 'unitary' coupling conductance in the region of 1nS in β-cells,[50] thereby accounting for almost all of the membrane conductance. Thus, we can reasonably predict that a near perfect space clamp arises within small clusters of pancreatic β-cells owing to their high input resistance and substantial gap junctional conductance *(equivalent to a low axial resistance in cable theory)*.

Taken together with the observations that distant deflection 'echoes' were not observed and that the uncompensated C_{slow} transient could be fitted to a single exponential in the form;

$$IC = Imin + Io.exp-\left(t - \frac{to}{\tau}\right)$$

---

[48] Eddlestone, G.T., Goncalves, A., Bangham, J.A., Rojas, E. (1984). Electrical coupling between cells in the islet of Langerhans from mouse. J.Memb.Biol. 77: 1-14.
[49] Andreu E. Soria B. Sanchez-Andres JV. (1997) Oscillation of gap junction electrical coupling in the mouse pancreatic islets of Langerhans. Journal of Physiology. 498 (Pt 3):753-61.
[50] Gopel, S., Kanno, T., Barg, S., Galvanovskis, J., Rorsmann, P. (1999) Voltage-gated and resting membrane currents recorded from B-cells in intact mouse pancreatic islets. J.Physiol. 521.3: 717-28.

where I_C is the capacitive current at time t; Imin is the whole-cell current to which the transient decays; Io is the initial capacitive current at t_o (zero time); and τ is the time constant for a first order decay given by Rs.Cm, where Rs is the access resistance and Cm is the membrane capacitance, the data are in empirical agreement with the contention that an effective space clamp had been achieved within the small cluster preparation at these more physiological temperatures.

From previous estimates of gap junctional conductance in acutely isolated, lone pairs of β-cells which indicated a coupling coefficient (CC) of 0.74 and a gap junctional conductance of 514 ± 137pS (G_j) within the 'active' phase[51], we can determine the potential of the 'driven cell' (V_2) from a simplified circuit which models the gap junctional resistance (Rgj) as being in series with the membrane resistance (R_m) of the second driven cell, a circuit which in effect takes the form of a simple 'voltage-divider'. Hence it might be argued that, under these conditions, the ratio of the potential of the driven cell (V_2) to the clamped cell (V_1) is given by;

$$\frac{V2}{V1} = CC = 0.74 = \frac{Rm}{Rgj + Rm}$$

where CC is the coupling coefficient. If Rgj is taken as 2GΩ (or 1/Ggj), then Rm will approach 6GΩ which is greater than the values obtained in our experiments. Taking the empirical observation that the central cell is effectively space-clamped and that the total input resistance of the small cluster was never observed to be less than 2GΩ, then it might be argued that the gap junctional resistance arising between the cell that we are recording from and a conjoined β-cell should amount to less than 5% of the sum of the input resistance of a single conjoined cell (taken as 5.7GΩ[51]) and the gap junctional resistance. Therefore, as no space clamp error is apparent, the gap junctional resistance might be expected to be < 300 MΩ (*i.e.* > 3.3nS).

Therefore, clusters maintained in primary culture may exhibit a gap junctional conductance which is at least six times greater than determined from measurements obtained from acutely-isolated cell pairs, a lower value which is potentially explained by the shearing damage that

---

[51] Andreu E. Soria B. Sanchez-Andres JV. (1997) Oscillation of gap junction electrical coupling in the mouse pancreatic islets of Langerhans. Journal of Physiology. 498 (Pt 3):753-61.

occurs during the mechanical trituration process which was employed to dissociate these cell pairs, and the greatly diminished rates of cellular metabolism which occur at room temperature.

Chapter Eight

The kinetics of ligand gated ion channels

Now we cross another conceptual bridge, as we enter the world of the ligand-gated ion channel. Like voltage-gated ion channels, they are usually rapidly activating and inactivating, usually within milliseconds, even in the continued presence of their ligand. This, of course, makes it very difficult to study their kinetics, given the technical issues arising in delivering a specific amount of neurotransmitter to a small patch with fractions of a millisecond.

Fortunately, nature has provided us with a somewhat more pedestrian model for ligand-gated ion channels in the $GABA_C$ receptor and, as luck would have it, Dr Jahanshah Amin had preliminary recordings from a variant of the ρ subunit which has turned this normally slow ligand-gated ion channel into one that took literally half an hour to open at low concentrations of GABA. Initially, I was asked to characterise the variant and we determined that the ultraslow kinetics could be pinpointed to a single naturally occurring mutation, or polymorphism at the Serine residue S439 which had been changed to a much bulkier phenylalanine residue (F). For the sake of brevity, we will not consider how Dr. Amin and his colleague Dr. Veronica Pollock cloned the variant, or how I used their constructed chimeras to identify the cause of the altered kinetics. Even more usefully, unlike other GABA receptors, functional $GABA_C$ class receptors can be formed from a pentameric arrangement of five identical ρ1 subunits, making interpretation of experiments in which we change a single amino acid via *site-directed mutagenesis* far easier to interpret.

$GABA_C$ receptor-channels are widely expressed within the interneurons of the central nervous system and putatively serve as autoreceptors through which the frequencies of neuronal networks are modulated. The ρ1 subunit contributes agonist (GABA) binding domains within the previously uncharacterized pentaoligomeric $GABA_C$ receptor complex. The isoform of the ρ1 subunit with greatly slowed kinetics of activation and deactivation in response to agonist was designated ρ1–in by Dr. Amin, which could be attributed to a single nucleotide polymorphism at S439F.

Now that we had identified this ultraslow variant S439F, I proceeded to create a spectrum of amino acid substitutions at ρ1$_{S439}$ via site-directed mutagenesis (see Appendix G) to probe the properties of this unique polymorphism further.

Classically studies of ligand-gated ion channels have been impeded by the rapidity with which agonists bind to and then gate (i.e. open or activate) such channels, the fast rate at which these ion channels then close upon dissociation of the agonist (channel closing or deactivation), and also by the rapid decline in channel activity during prolonged exposure to agonist (a process known as desensitisation). At the single channel level, transitions between open, closed and desensitized states are therefore very difficult to discriminate owing to their sub-millisecond time scale, by the superimposition of rapid and concurrent transitions between open and desensitized channel states, and by the expression of receptor mosaics of variable subunit composition. Although many ligand-gated ion channels are voltage-sensitive, we are spared yet another layer of complexity with the ρ1 receptor channel, which exhibits little or no voltage sensitivity.

The expression of ρ1 receptors composed of five identical subunits greatly simplifies our understanding of the fundamental mechanisms governing the kinetics of binding and gating of ligand-gated ion channels, not only because they are significantly slower, but also because they display little or no desensitization upon the prolonged application of high agonist concentrations, effectively removing the complexity of desensitised channel states from any kinetic analysis of channel activation and deactivation.

Thus, the use of homo-oligomeric ρ1 receptor-channels considerably simplifies matters and enables us to build a clearer model of the relationship between the primary structure of a ligand-gated receptor-channel (*i.e. its amino acid sequence*) and its kinetic properties. So far we know that specific residues present within the N-terminus of the receptor are responsible for agonist binding in GABA$_A$ and ρ1 homo-oligomeric receptor channels courtesy of Dr Amin and Professor Weiss, and that specific residues within the second transmembrane domain (TM2) contribute to what we believe is the ion channel pore, the selectivity filter which determines anion specificity (by forming a ring of positively charged side chains) and the transduction of agonist binding into channel gating. Further, residues within TM2 (Walters et al., 2000) and TM3 (Amin, 1999) determine the sensitivity of GABA receptors to anaesthetics and barbiturates, respectively. However, up until this point, no functional role had been elucidated for any residue within the proposed fourth transmembrane spanning domain.

Previously Amin & Weiss (1996) empirically determined that there was a minimum requirement for three agonist molecules to bind to gate (i.e. open) the homo-oligomeric $\rho 1$ receptor-channel, which would correspond to a minimum of three channel closed states *(we will address this subsequently)*.

Amino acid substitution at position S439 produces variable channel kinetics

As discussed, the expression of the original $\rho_{1\text{-in}}$ clone resulted in a receptor-channel with dramatically slowed kinetics of activation at low agonist concentrations and slowed kinetics of deactivation following agonist removal. Substitution at residue 439 with a spectrum of amino acids of varying size charge and hydrophobicity resulted in the expression of receptors with greatly variable kinetics of activation and deactivation in *Xenopus* oocytes, as measured by determining the $t_{1/2}$ for both parameters (*i.e.* the time for half-maximal activation/inactivation). This somewhat more descriptive kinetic analysis takes the form of measuring the time taken to achieve a half-maximal current deflection from the current baseline prior to application of the agonist (the $t_{1/2}$ of activation) or closure from full steady-state activation (*i.e.* an equilibrated agonist-receptor state; *i.e.* $t_{1/2}$ of deactivation).

However, not every substitution produced functional ion channels, as the expression of cRNAs for $\rho 1$ subunits with a substitution of Serine at position 439 for amino acids containing bulky hydrophobic residues, specifically Isoleucine, Tryptophan, and Tyrosine did not respond to any concentration of GABA tested *(summarised in the table below)*.

Point mutation	I_{max} (nA ± SE)	Hill Coefficient (n)	EC_{50} (μM ± SE)	n
$\rho1_{S439}$	930±174	2.36±0.09	0.78±0.05	10
$\rho1_{S439A}$	1057±169	2.80±0.24	0.69±0.10	9
$\rho1_{S439G}$	195±61	2.50±0.08	0.60±0.01	4
$\rho1_{S439M}$	305±141	1.60±0.14	0.34±0.03	8
$\rho1_{S439F}$	232±81	1.62±0.17	0.37±0.05	4
$\rho1_{S439T}$	40±4	2.05±0.08	0.35±0.02	6
$\rho1_{Y438W}$	664±142	2.76±0.19	0.57±0.02	3
$\rho1_{R440K}$	423±182	2.15±0.01	1.26±0.01	4
$\rho1_{R440A}$	348±232	3.03±0.09	0.47±0.01	4
$\rho1_{S439W}^{\chi}$, $\rho1_{S439I}^{\chi}$ $\rho1_{S439Y}^{\chi}$				

Table 1. variation in the maximal amplitude of the current evoked, half-maximal effective concentration (EC_{50}), and Hill coefficient values (a measurement of the degree of cooperativity occurring during the activation process instigated through ligand binding)[52] from the expression of equivalent amounts of cRNA of ρ_1 for all amino acid substitutions (ρ_{1S439X}) which were created by site-directed mutagenesis (χ denotes that the construct did not produce functional channel or leak current).

Classically, these values are derived from a dose-response relationship which is generated by testing the response of a system (in this case the size of the current evoked) across a range of concentrations of agonist (or inhibitor) spanning many orders of magnitude.

These dose-response relationships, in this case current amplitudes vs. the concentration of the agonist GABA, or {A}, were fitted to the normalized form of the Hill equation in the form;

---

[52] Classically used in the study of haemoglobin

$$I = I_{max} / (1+[EC_{50}/\{A\}]^n)$$

Where **I** is the peak current evoked after application of agonist at concentration $\{A\}$; $\mathbf{I_{max}}$ is the peak current evoked at saturating agonist concentrations, the $\mathbf{EC_{50}}$ is the agonist concentration at which a half-maximal current response is evoked and *n* is the Hill coefficient. Data were fitted to the Hill equation after first normalizing the data to the extrapolated maximum value ($\mathbf{I_{max}}$), the means of the data are presented ± SE as a function of agonist concentration for the number of experiments given.

For those amino substitutions which did form functional channels, superimposed traces of agonist-evoked currents are given across a broad µM concentration range in the figure below (panel A), and the kinetic description of the data is summarized by presenting $t_{1/2}$ measurements (mean ± SE) for both activation (B) and deactivation (C) as a function of GABA concentration normalised as a ratio of the pre-determined EC_{50} derived for each amino substitution. The $t_{1/2}$ values derived for both channel activation and deactivation increased in the order (slowest to fastest) Phenylalanine > Methionine ≥ Threonine > Glycine > Alanine ≥ Serine. At low GABA concentrations, ρ_{S439F} was a full order of magnitude slower in activation at an equivalent GABA concentration than either wild-type receptor-channel (designated ρ_{S439}) or ρ_{S439A} (B). As you may observe, at saturating agonist concentrations, the rate of channel activation was almost indistinguishable.

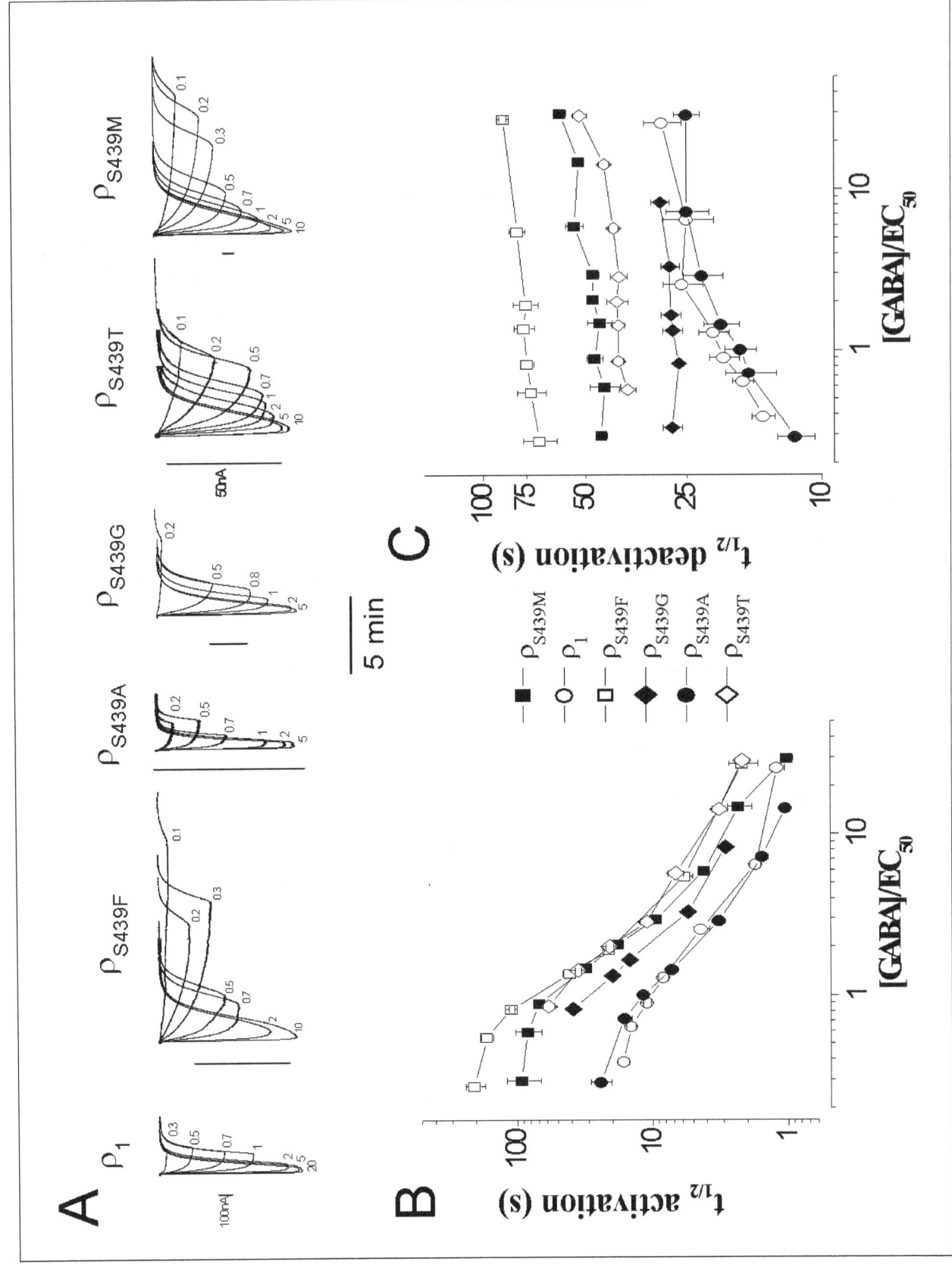

Effect of amino acid substitutions at position 439 upon the kinetics of channel activation and deactivation. Panel A. Families of currents evoked by application of GABA across a concentration range are presented superimposed for wild-type ρ1 (ρ$_{S439}$) and for each amino acid substitution that gave rise to a functional channel. Current deflections elicited by GABA at a holding potential of −70mV are shown across the μM concentration range indicated to the right of each evoked current trace for (wt) ρ$_{S439}$, ρ$_{S439F}$, ρ$_{S439A}$, ρ$_{S439G}$, ρ$_{S439T}$, and ρ$_{S439M}$. Current traces are presented on the same time scale and normalized to the point of agonist application. Vertical scale bars represent 100nA for all residues except ρ$_{S439T}$ (50nM) as indicated. (B & C). The time taken for a 50% change in the maximal GABA-evoked current deflection amplitude ($t_{1/2}$) is presented as a function of GABA concentration normalized to the determined EC50 value for each substitution for both receptor-channel activation (panel B) and (panel C) deactivation for n values (means ± SE).

The argument that these mutations result in a slowing of the rate of channel opening ('gating') following agonist binding is effectively dispelled by the observation that the slow activation seen at low agonist concentrations is dramatically overcome at higher agonist concentrations.

This observation might be taken to infer that the slow kinetics might rather result from either slower agonist binding, or else a slowing in the rate at which conformational changes induced by agonist binding are mechanistically 'reported' or presented to the channel gating mechanism. Thus, we may infer that the slow kinetics of activation of these substitutions allude to a 'conformational-binding' effect rather than a direct impediment to channel gating itself. A similar profile in the slowing of the kinetics of channel deactivation, although differences in both the concentration-dependence and magnitude of changes in the rate of channel deactivation were far less marked than for activation. Note that the $t_{1/2}$ of deactivation showed little concentration-dependence (except for Alanine and Serine).

Substitution at 439 produces only small changes in apparent affinity

Current amplitudes evoked at each agonist concentration tested were plotted as a function of the *normalised* maximal current evoked (which was determined by extrapolation upon fitting to the Hill equation). As the homo-oligomeric ρ1 receptor shows only minimal desensitization upon prolonged agonist application at higher concentrations, fits to the Hill equation are therefore not distorted by desensitisation. Parameters derived from a fit to the Hill equation indicate that the apparent affinities for all substitutions lie within an order of magnitude (see table 1) despite a dramatic variation in channel kinetics, although the Hill coefficient appears to imply significantly more *co-operativity* for Methionine and Phenylalanine than for the other smaller, less hydrophobic residues.

The findings suggest that S439 does not contribute substantially to determining either the apparent affinity or cooperativity of the action of agonist in channel gating (*i.e.* opening and closing), and therefore the greatly slowed rates of receptor-channel activation and inactivation observed for ρ_{S439F} and ρ_{S439M} do not appear to derive from dramatic changes in the apparent affinity of the receptor-channel for its agonist. Further, the absence of any significant effect of ρ_{S439A} upon any of the parameters tested strongly suggests that the differences in kinetic rates observed between $\rho 1_{S439}$ and $\rho 1_{S439F}$ do not arise due to the loss of a functional phosphorylation site.

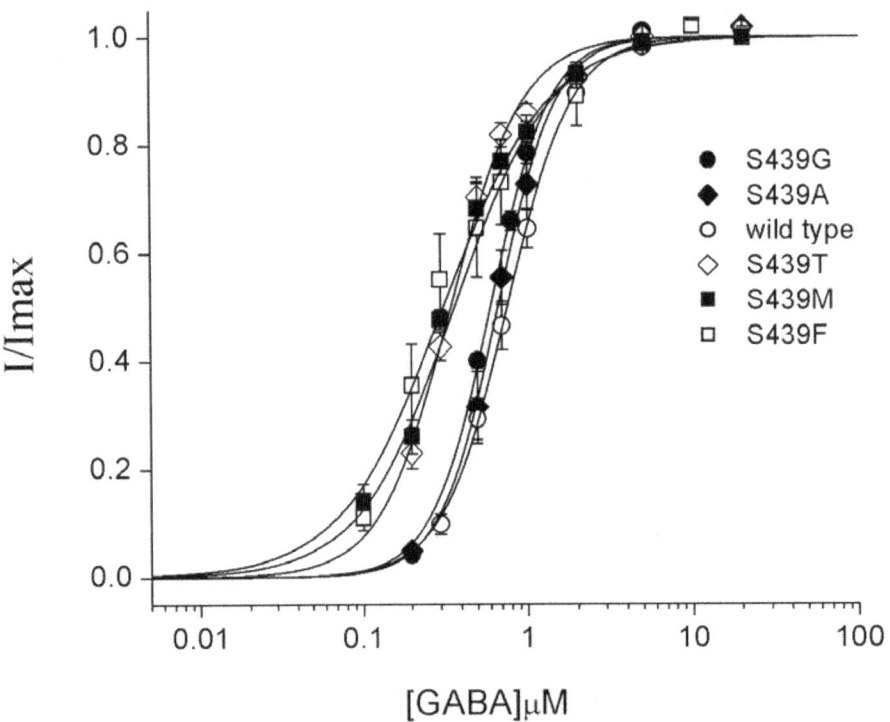

The graph presented above shows the concentration-response relations for the wild-type $\rho 1$ receptor, ρ_{S439F}, ρ_{S439A}, ρ_{S439G}, ρ_{S439T} and ρ_{S439M}, where peak current deflection amplitudes are plotted as a function of GABA concentration (given as µM) after normalization to the extrapolated maximal agonist response predicted by a fit to each data set to the Hill equation *(see methods)*. The means ± SE for each agonist concentration tested are presented together with a best fit to the Hill equation *(the associated numerical data are presented in Table 1)*.

Effects of amino acid substitution at S439 upon mean current amplitudes

Consistently lower levels of current amplitude were observed upon substitution of S439 with other amino acids, most notably *Threonine* (T), *Phenylalanine* (F) and *Methionine* (M; see Table 1). This may variously be explained by a reduction in mean open probability (P_o) at a given effective agonist concentration; a reduction in the number of functional channels (n); or a decrease in unitary channel conductance (i) due to a reduced anion permeation rate, as the macroscopic current is given by the term $I = i.n.P_o$. The apparent inability of other large hydrophobic residues (*i.e.* ρ_{S439I}, ρ_{S439W} or ρ_{S439Y}) to form functional ligand-gated channels upon expression in *Xenopus* oocytes implies that a very tightly restricted binding pocket may be formed within the quaternary structure of the assembled pentameric channel complex.

Substitution at adjacent residues has little influence upon channel kinetics

Site-directed mutagenesis at adjacent residues to S439 was performed to elucidate whether the dramatically slowed kinetics following substitution of bulky hydrophilic residues at this position was a phenomenon specific to a certain residue, or else could be attributed to changes in the hydrophobicity within the general vicinity of a domain. Although some substitutions at Y438 and R440 resulted in receptor-channels that were not functionally expressed, the agonist-evoked current traces elicited for the functionally expressed channels Y438W, R440A and R440K are presented in the following figure. However, none of these substitutions reproduced the dramatically slowed kinetics of activation which were characteristic of ρ_{S439F}.

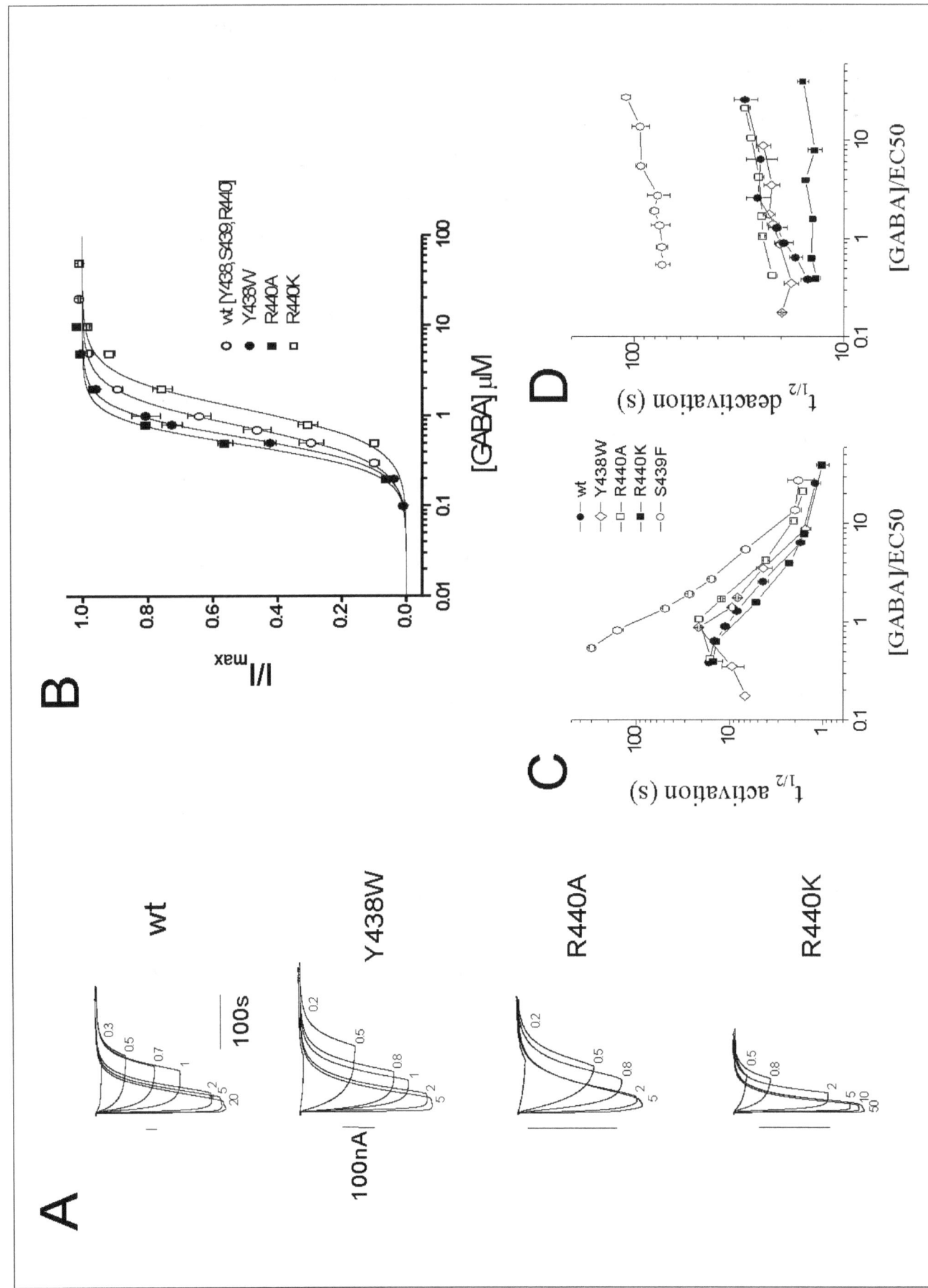

Substitution at Y438 or R440 does not dramatically slow ρ1 receptor kinetics. **Panel A** shows the current deflections elicited by GABA at a holding potential of –70mV across the range of concentrations (μM) indicated to the right of the respective current traces shown for wild-type ρ1 receptor-channel and for Y438W, R440A, and R440K. Current traces are presented on the same time scale and normalized to the point of agonist application. Vertical scale bars represent 100nA for all current families shown. **Panel B** shows the concentration-response relations for the wild-type ρ1, Y438W, R440A and R440K receptor-channels, wherein peak current deflection amplitudes are plotted as a function of GABA concentration (given as μM) after normalization to the extrapolated maximal agonist response predicted by a fit of each data set with the Hill equation (see methods). The means (± SE) for each agonist concentration tested are presented together with the best fit to the Hill equation and the associated numerical data are presented in Table 1. The time taken to reach 50% of the maximal GABA-evoked current deflection amplitude ($t_{1/2}$) is presented as a function of the agonist concentration normalized to the EC50 value for each substitution shown above for receptor-channel activation (**Panel C**) and deactivation (**Panel D**) with the values for the S439F mutant also presented on a logarithmic scale for comparison. Data are presented as means ± SE.

Measurements of $t_{1/2}$ values for receptor-channel activation for substitutions at Y438 and R440 are plotted as a function of the equivalent effective concentration of GABA (see panel C above) on a logarithmic scale, with the data contrasted against $\rho 1_{S439F}$ and $\rho 1_{S439}$. Point substitutions at Y438 and R440 within the *YSR* motif (amino acid residues 438-440) yielded only a small variation in the derived Hill coefficient and EC_{50} parameters (see panel B above and Table 1). The rates of receptor-channel deactivation are similarly shown as a function of equivalent effective GABA concentrations in panel D.

Two kinetic changes induced by substitution at adjacent positions are however noteworthy. Y438W exhibits a prominent biphasic concentration dependence in the $t_{1/2}$ of activation, the rate first increasing and then decreasing as a function of agonist concentration (panel C), although values were otherwise comparable to wt ρ1 control values. This may somehow relate to tryptophan's amphipathic properties at the residue's putative position at the interface between the lipophilic and hydrophilic regions of the plasma membrane.

Further, the $t_{1/2}$ values for R440K receptor-channel deactivation, considering that this was intended as a conservative amino acid substitution, are *significantly faster* than for the wt, R440A, Y438W channels or for the spectrum of changes introduced at S439 (see panel D above). The failure of other non-positively charged substitutions to produce functional channels at this position and the differences in deactivation rates between R440, R440A, and R440K suggest *that this positively charged amino residue plays a role in determining the gating kinetics of deactivation*. This may, however, reflect the channel's lower affinity for GABA (Panel B) which might also be expected to result in a more rapid dissociation of agonist from the channel, also it serves to further emphasize the critical importance of the stereochemistry contributed by the side chains present within this YSR motif at the interface of the putative fourth transmembrane spanning domain of the GABA receptor superfamily.

The kinetics of channel deactivation change with agonist concentration

The classical model of ρ1 receptor-channel gating postulated by Amin & Weiss[53] predicted three closed states and one open state which was later extended to two open states by measurement of [^{3}H] GABA dissociation rates, one of which was determined to arise from the effective 'locking' of agonist onto the receptor-channel upon opening.[54] The rate of ρ1 receptor-channel deactivation, in contrast to activation, is demonstrably slowed with increasing agonist concentration. Initially, fitting was performed according to the original model of Amin & Weiss which predicted that the decay in the current upon agonist dissociation would be described by a *single open state* and thus fitted to *a single exponential*. However, at higher agonist concentrations (please see figure below), neither the wild-type ρ1 receptor-channel, nor any of the receptor-channels with amino acid substitutions at S439 could be satisfactorily described by a single exponential, a phenomenon that was particularly evident for the ρ_{S439F} and ρ_{S439M} mutants which also exhibit greatly slowed kinetics of deactivation.

---

[53] Amin, J., Weiss, D.S. Insights into the activation of ρ1 GABA receptors obtained by co-expression of wild-type and activation-impaired subunits. Proc. R. Soc. Lond. Ser. B Biol. Sci. 263: 273-82 (1996)
[54] Chang, Y & Weiss D.S. (1999). Channel opening locks agonist onto the GABAC receptor. Nature Neuroscience 2: 219-225.

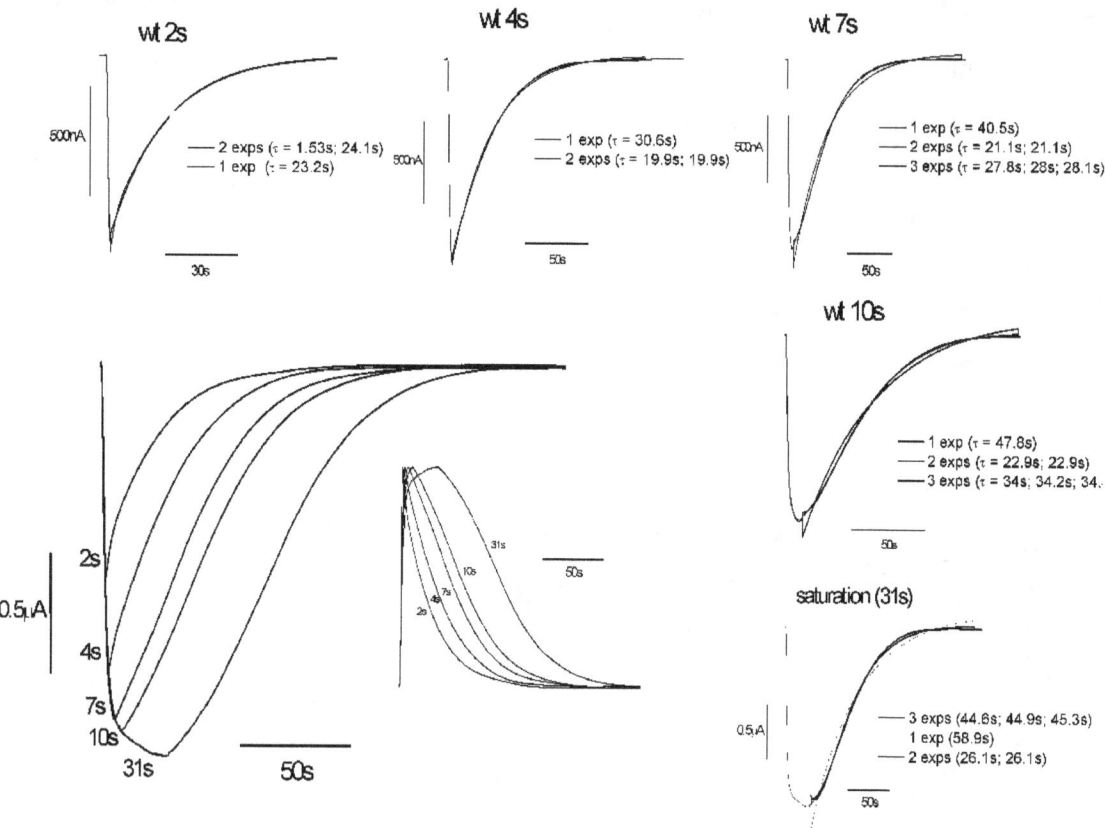

Fast perfusion reveals a time-dependence in the number of open states occupied

The figure above shows the kinetics of the time course of deactivation at near saturating concentrations of GABA for the wt $\rho 1$ receptor channel which is described by a fit to a minimum of three, but not one or two exponentials using a *fast valve-driven perfusion system*. A concentration of twice the pre-determined EC50 for GABA is applied for progressively longer time intervals of 2, 4, 7, 10 and 31 seconds until current saturation is achieved.

At low agonist concentrations, the kinetic data is in apparent agreement with a single open state model as the current decay upon agonist removal *is satisfactorily described a single exponential* for the wild-type $\rho 1$ receptor-channel. Yet at intermediate GABA concentrations, a good description of the deactivation of both the $\rho 1$ and $\rho 1_{S439F}$ receptor-channels is given by a fit to two exponentials, suggestive of the existence of a receptor population predominantly with two-open states occupied at equilibrium. The occupation of a third open state towards saturating concentrations is consistent with the view that the fourth and fifth binding sites of the pentameric ρ1 receptor-channel complex are indeed functional, and that their occupancy is driven in a concentration-dependent manner. This phenomenon is clearer for the ρ1$_{S439F}$ receptor-channel variant as shown below.

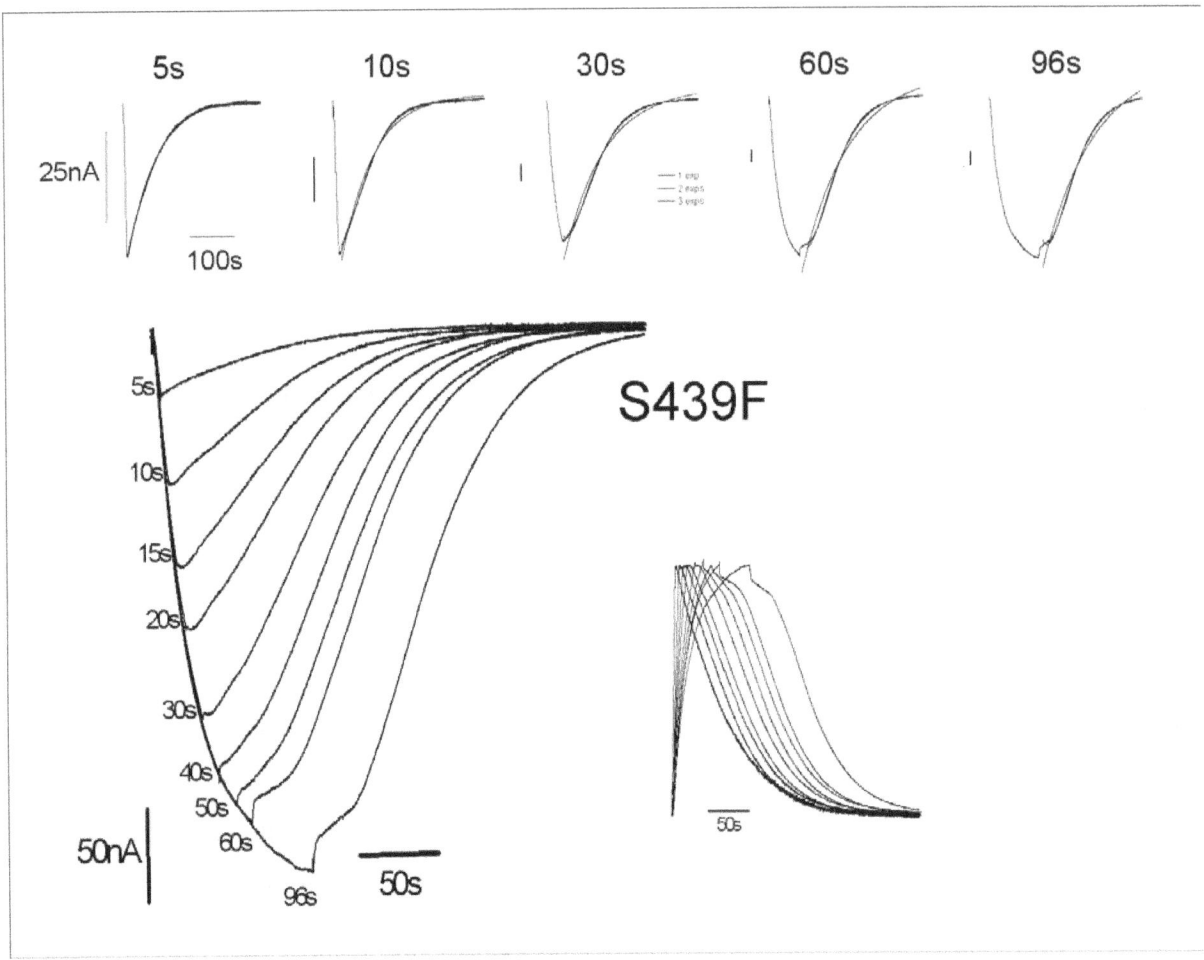

If the mean number of open states occupied, a reflection of the number of agonist molecules functionally bound, is seen to increase in a concentration-dependent manner then, due to the inherently slow nature of ρ1 receptor-channel activation, the number of bound states might also be expected to increase in a time-dependent manner at these higher agonist concentrations where multiple exponential fits to channel activation are observed. The apparent description of the sigmoidal pattern of ρ1 and ρ1$_{S439F}$ receptor-channel deactivation by three exponentials at higher agonist concentrations might however in part be attributed to some manner of artifact generated by a slow perfusion system. To address both these issues, a faster perfusion method was realized by a parallel port application of solutions directly into the chamber using solenoid valves to switch instantaneously between solutions effectively eliminating dead space. Thus the 99% exchange time for agonist removal was reduced to less than 2s, with a chamber $t_{1/2}$ of 0.31s. To specifically test the hypothesis that the number of open states occupied increased with the duration of agonist application, the chamber was alternately perfused with agonist-free OR_2 media, then with agonist at twice the pre-determined EC_{50} concentration for a fixed time interval, and finally with agonist-free OR_2 to determine the rate of receptor-channel deactivation.

The preceding figures illustrate the marked time-dependence of the kinetics of $\rho 1$ and $\rho 1_{S439F}$ receptor-channel deactivation following agonist washout which was quantified by the fitting of exponential time courses immediately subsequent to the point of solution change, which was conveniently marked by the small electrical stimulus artifact caused by the switching of solutions. The residuals describing the quality of the exponential fits are presented in table 2 below, whilst the mean values for the 1st, 2nd and 3rd time constants ($\tau_{1..3}$) derived for the pre-determined number of open states (exponentials) which best described the time course of receptor-channel deactivation are given in table 3.

Time of application (s)	Single exponential	Two exponentials	Three exponentials
wt			
2	0.9922	0.9992	--
4	0.9975	0.9976	0.9979
7	0.9892	0.9922	0.9991
10	0.9814	0.9903	0.9993
31	0.9688	0.9809	0.998
S439F			
5	0.997	0.9982	--
10	0.9926	0.9924	0.9961
15	0.9912	0.9931	0.9993
20	0.989	0.9994	0.9993
30	0.9861	0.9992	0.9992
40	0.9815	0.9815	0.9988
50	0.9793	0.9889	0.9992
60	0.9758	0.9989	0.999
96	0.9792	0.9987	0.999

Table 2. Residuals obtained for fits to one, two and three exponentials for the wt and ρ1S439F receptor channels for different durations of agonist application at twice the effective EC_{50} concentration.

ρS439F	Mean	± SE	ρ1	mean	± SE
5s τ_1	58.2	3.32	2s τ_1	11.2	5.39
10s τ_1	34.4	0.282	2s τ_2	22.2	0.95
10s τ_2	34.4	0.282	4s τ_1	21.4	0.76
15s τ_1	24.0	0.392	4s τ_2	21.4	0.76
15s τ_2	24.1	0.442	7s τ_1	30.3	5.35
15s τ_3	25.6	1.27	7s τ_2	30.5	5.38
20s τ_1	31.9	6.87	7s τ_3	30.6	5.40
20s τ_2	32.0	6.20	10s τ_1	40.1	4.31
20s τ_3	37.0	4.42	10s τ_2	40.4	6.15
30s τ_1	37.2	5.45	10s τ_3	40.6	6.15
30s τ_2	37.4	5.46	31s* τ_1	46.5	5.86
30s τ_3	37.6	5.50	31s* τ_2	46.5	4.05
40s τ_1	42.8	5.89	31s* τ_3	46.9	4.15
40s τ_2	43.0	6.65			
40s τ_3	44.2	5.68			
60s τ_1	45.6	9.27			
60s τ_2	45.8	9.36			
60s τ_3	49.0	7.05			
96s* τ_1	39.9	11.2			
96s* τ_2	40.0	7.96			
96s* τ_3	45.0	3.84			

Table 3. Mean values obtained for the 1st, 2nd and 3rd time constants ($\tau_{1..3}$) derived from the pre-determined number of open states (exponentials) which best described the time course of receptor-channel deactivation (table 2). Determination of the number of open states occupied after different exposure times to agonist concentrations that were twice the EC_{50} for the recordings illustrated previously using the non-linear least squares fitting algorithm (Origin v.6.0). Unit-less residuals are given for fits to one, two and three exponentials of freely varying starting amplitude. Quality of exponential fits to the recordings obtained were determined by observation of the correlation of the superimposition of the exponential best fits upon the current traces and by comparing them with the residual values obtained. * denotes the time point at which receptor-current saturated.

Thus, as the duration of agonist application declined, so the rate of channel deactivation was seen to increase, its time course approaching that described by a single exponential for the briefest agonist applications (5 s and 2s application for ρ_{S439F} and ρ_1, respectively). However, as the duration of agonist application is increased, so the time course of decay becomes sigmoidal consistent with fits to multiple exponentials of differing starting amplitudes for both ρ_1 and ρ_{S439F} receptor-channels. This suggests the existence of multiple channel open states and adds support for the existence of a 'locked' channel open state(s) as proposed by Chang and Weiss. For intermediate application times, the decay was most simply described by two exponentials for both ρ_1 (4s) and $\rho_{1\text{-}S439F}$ (10s), suggesting a temporal and concentration-dependent hierarchy of sequential state transitions and that the fundamental properties of the ρ_1 and ρ_{S439F} receptor-channels are not changed by the S439 polymorphism, rather the rate at which conformational transitions upon agonist binding occur.

However, upon closer inspection of the fits to 2 and 3 exponentials (table 3) gave rise to time constants of similar magnitude, but with decremental amplitudes. This observation suggests the existence of kinetic events (steps) with equivalent rate constants but which are sequentially realized by virtue of the requirement for one open state to be vacated before another can start to do so which, in a sense, constitutes a form of locking mechanism. It is, therefore, logical to ascribe the sigmoidal current decay at higher concentrations to a sequential progression through three channel open states, all of which are in turn realized in a time and concentration-dependent manner as demonstrated. This is sufficient to account for the sigmoidal nature of the current deactivation, the temporal separation of the exponentials expressed in terms of different starting amplitudes.

Intriguingly however, the $\rho1_{S439F}$ receptor-channel also displayed a further rapid state transition not readily apparent within the wild-type $\rho1_{S439}$ receptor-channel deactivation, which was very rapid (1.83±0.3s, n=5 after current saturation) and time-dependent in both its onset and magnitude, and which was fully reversible (and present) in all-time course recordings, possibly indicative of a further 4th channel open state which might serve to explain the slow saturation of these receptors, one which is characterized by a very small forward opening rate constant

($\beta \leq 1 \times 10^{-3}$ s^{-1}), and a much faster reverse closing rate constant ($\alpha \leq 1$s^{-1}), where $\tau = 1/k$, and where k (α or β) is the rate constant.

Correlations between residue properties and measured kinetic parameters

The established values for molal volume and hydropathicity index, bulkiness, polarity, pI, pK and hydrophobicity, molar volume, solvent parameter index, and hydrophilicity were all plotted in turn as a function of the parameters derived from our fits to the Hill equation and to the $t_{1/2}$ values derived from the six functional substitutions derived at position 439.

Fits to the $t_{1/2}$ values obtained show that the $t_{1/2}$ of receptor-channel activation correlates strongly with both the molar volume and hydrophobicity, but most strongly with the hydrophobicity product (*i.e.* the product of the molar volume and hydrophobicity, a parameter which otherwise explains the otherwise paradoxical result observed upon glycine substitution. In contrast, $t_{1/2}$ values for receptor-channel deactivation correlated most strongly with the molar volume of the side chain, again reinforcing the idea that receptor activation and deactivation are not simply thermodynamic and kinetic reversals of the same given process, as is also revealed by differences in the time and concentration dependence of kinetics of activation and deactivation.

Despite the narrow range of values obtained from the fitting of current dose-response relationships to the Hill equation, significant correlations were obtained for both the derived Hill coefficients and EC$_{50}$ values when plotted as a function of molar volume. The range of Hill coefficients, regarded as a measure of the degree of cooperativity in the activation of an enzymatic process by a ligand, in this case an ion channel, most strongly correlated with the molar volume ($P<0.005$) although they also correlated more weakly with the hydrophobicity via linear regression analysis.

Conclusions

The cloning of a single nucleotide polymorphism within the $\rho 1$ subunit, which contributes residues towards the agonist binding domain of the penta-oligomeric GABA$_C$ receptor complex, has provided us with a powerful model through which we can obtain a clearer understanding of the gating kinetics of ligand-gated ion channels. The greatly slowed kinetics of the $\rho1_{S439F}$ receptor-channel variant enables us to study the relationship between structure and function in real time.

Both the rates of $\rho1$ receptor-channel activation and deactivation, in addition to the Hill coefficient, were shown to correlate strongly with both the molecular volume and the hydrophobicity of the residue at position 439. The molecular volume argument is especially compelling in the light of the slowed kinetics of the S439T substitution. The large differences in the rates of activation between the fastest ($\rho1$) and the slowest ($\rho1_{S439F}$) substitutions could not however be attributed to changes in the rate of channel gating, apparent receptor affinities, or in the number of open states occupied, but arose either due to slower rates of agonist binding or to a retardation of the resulting conformational changes elicited by agonist binding, or both.

The kinetic and Hill equation parameters derived from mutagenesis within the *YSR motif* lend further credence to the argument that the apparent affinity of agonist binding, the conformational changes that are induced as a result of agonist binding, and the rates of channel gating are all somehow kinetically independent and separable events that do not change in parallel. The presence of a phenylalanine at position 439 indicates that elements within the fourth transmembrane spanning domain govern channel kinetics, although whilst the rate of channel opening (gating) does not appear to be rate-limiting at high agonist concentrations, the mutation appears to indicate that rate constants for agonist binding and the resultant conformational changes that are induced constitute a separable process from those of channel gating. Moreover, the slow time course of activation and deactivation, the differing amplitudes obtained from exponential fitting to channel deactivation, and Hill coefficients derived from fits to concentration-current response relationships strongly suggest that agonist binding is a highly cooperative process.

It has been proposed that the conserved sequences present within related receptor super-families reveal not only the key binding domains and catalytic sites, but also predict the trail of hydrophobic residues present within the cores of proteins that serve as conduits for the transmission of conformational information from one interface of agonist-protein or protein-

protein interaction to another more distant site. Thus, these hydrophobic pathways serve to effect the propagation of information concerning the activity of a protein, for example, that information that is involved in the allosteric or ligand-mediated activation or modulation of protein activities. Indeed, a disruption of these hydrophobic pathways involved in the conformational coupling of information between distant interfaces and active sites by mutagenesis may stabilise either the active or inactive form of a protein. It is possible that the creation of a hydrophobic binding pocket within $\rho1_{S439F}$ may lead to a change in the stability of a pathway that conformationally couples ligand binding events in one domain to those of another, thus contributing to the slow kinetics of $\rho1_{S439F}$ observed at low agonist concentrations.

The extent of $\rho1$ receptor-channel activation (I/I_{max}) could be shown to result from the sequential co-operative occupation of multiple agonist binding sites in a time- and concentration-dependent manner, due to the requirement for multiple exponentials with differing starting amplitudes to describe the deactivation process. The number of open states realized, which are achieved in a concentration and time-dependent manner, will thus determine the extent of channel activation and alter the rate of channel deactivation (which was more rapid for lower agonist concentrations), as receptor-channel deactivation also represents a sequential process of state transitions.

The slow kinetics of the mutant receptors might be attributable to a slowing in the rate constants of those co-operative conformational changes that are induced by agonist binding, rather than by a slowed channel gating mechanism, as the slow kinetics of activation are overcome at higher agonist concentrations. However, as the decay of the current during deactivation represents the relaxation of the GABA receptor from its ligand bound to unbound states, this decay (deactivation) must be dictated primarily by the rate of ligand unbinding, as channel gating transitions from open to closed channel states are known to occur within the millisecond to microsecond timescale as opposed to the seconds to minutes taken for $\rho1$ receptor-channel deactivation. An increase in apparent agonist affinity may in part help to explain why channel deactivation is not explained by a single exponential for the $\rho1_{S439F}$ receptor-channel at lower agonist concentrations, as channel deactivation is dependent upon agonist affinity, whereas gating is not. Intriguingly, differences in the energy barrier between the unbound and bound states serve to define both agonist affinity and the deactivation rate, which might serve to explain the relatively smaller differences in the kinetic rates of deactivation and apparent receptor affinity for the wild-type $\rho 1$ and ρ_{S439F} relative to the larger differences in kinetic rates of activation observed at lower agonist concentrations.

The much slower $\rho 1$ receptor-channel mutants reveal additional kinetic complexities arising from the co-operative binding of agonist to five subunits and the resulting sequential communication of these binding events to the gating mechanism via conformational changes to maximally activate the receptor, a gating event which cannot itself be the rate-limiting event, as at saturating agonist concentrations little difference is seen between the $t_{1/2}$ of receptor-channel activation between the wt $\rho 1$ and ρ_{S439F} receptor-channels. Such a distinction between gating and binding-conformational events is thus made possible by the $\rho 1_{S439F}$ receptor-channel polymorphism. It is clearly apparent that the number of open states increases in a concentration-dependent manner, suggesting that highly bound states are thermodynamically favoured by mass action at higher agonist concentrations.

In summary, just as the number of closed states of an ion channel may be determined by the number of exponentials required to describe the kinetics of its activation, so the number of exponentials required to describe the kinetics of ion channel inactivation indicates the number of channel open states. As we have determined, the number of open states occupied by the $\rho 1$ receptor-channel is dependent upon both the agonist concentration and the duration of its application. However, for the $\rho 1$ receptor-channel, a minimum of a three open state kinetic model is required to explain its behavior, with the possibility of a fourth open state which locks the agonist onto the receptor.

Chapter Nine

Final reflections

This book is unorthodox in that it charts a conceptual and experiential approach to ion channel kinetics. In fact, all the experiments published within the book have been presented in their chronological order and, save for some of the formative recordings presented in Chapters Two and Three, none have previously been published other than as conference proceedings.

My entry into the world of biophysics was classical in that I was left to my own devices in a small corner room with a box of old electronic components and I spent much of the first few months of my Ph.D. studentship wandering across the Babraham Institute begging and borrowing equipment to construct my own electrophysiological rig from first principles. Much the same applied to my understanding of biophysical theory as I only had access to a limited selection of periodicals and the option to buy a set of classical works which I duly ordered, read and attempted to memorise.

Thus, it is intended that this work should lower the conceptual barrier to entry into the realm of biophysics by approaching biophysical solutions through the portal of 'real' scientific problems, as understanding classical biophysical theory from a purely abstract perspective is an insurmountable challenge to many.

However, once you have successfully navigated this introductory work, the author wholeheartedly recommends that you then revert to the classical works of Hille,[55] Sakmann & Neher[56] and Aidley & Stanfield,[57] *inter alia*.

Conventionally, all the authors who laid the foundations of this research, as well as those who have followed on during the decades which have elapsed since it was performed, may well feel that they should have been cited throughout this work. However, as the total number of citations would have literally run into the many thousands, the author can only beg the patience and forgiveness of those who feel they have been overlooked, given that recognition for one's

---

[55] Ion channels of excitable membranes, Bertil Hille, 3rd edition, Sinauer, 2001.
[56] Single channel recording, Bert Sakmann & Erwin Neher, Springer, 1983.
[57] Ion Channels: Molecules in Action, David J. Aidley & Peter R. Stanfield, Cambridge University Press, 1996.

work is the only reward that most scientists ever receive. Therefore, owing to spatial limitation and political considerations, this compendium has erred on the side of caution and has only cited key sources rather than being selective in who should be included or excluded.

If anything stands to be achieved by this work, it is the hope that the 'cognitive barrier' precluding access to this complex field has been lessened a little, thereby affording others a precious insight into the elemental nature of electricity within living systems. Upon occasion, the properties of ion channels appear miraculous, while at other times they seem bewildering if not perplexing. One should, however, bear in mind that the solutions nature has engineered over hundreds of millions of years of trial and error are often elegant in their simplicity.

To illustrate this point, I shall refer to one of my favourite scientific findings, in which various authors probed the weighting of synaptic inputs arising within the dendritic tree at various distances from the soma, and to their surprise found that the answer was always 'one' as the more distal inputs, facing a greater electrotonic decay, compensated for their weaker voice by having a greater synaptic weighting, or by 'shouting more loudly'.[58]

Thus, it seems that simple models and enduring principles can still be derived from wondrous complexity.

---

[58] Neuroscience. Synapses shout to overcome distance. Laura Helmuth. Science. 2000 Aug 25;289(5483):1273.

ABOUT THE AUTHOR

The author was educated in Northamptonshire, Cardiff, and Cambridge, descending a long line of medical doctors and the first Welshman to captain the English cricket team, Cyril F. Walters. A keen sportsman, he played rugby for St. Edmund's College and New York, but was perhaps better known as a martial artist. After completing his Ph.D. in Cambridge, he underwent his post-doctoral training in New York, Brussels, Durham & Tampa, before being appointed as a group leader in a joint position between the Universities of Dusseldorf and Los Angeles which, regrettably, eventually fell through…

Although there is enough unpublished data from 12 years in the laboratory to write further books on second messengers and the pharmacology of ion channels, this work focuses exclusively on his as yet unpublished works on ion channel kinetics so that the knowledge which was gleaned at such a high cost is not forever lost to the scientific community.

The author is currently an active editor and photographer based in London, the city of his birth.

APPENDIX A

Whole cell recording methods for small intestinal crypts

Isolation of viable crypts suitable for patch-clamp electrophysiology

The initial stage of preparation of intact crypts was performed as described previously. Fractions were collected over timed 10, 6, 2, 2 and 2-minute intervals. The three fractions from minutes 17-22 were centrifuged at 50g for 1 minute in a Super Minor MSE centrifuge and washed in Hanks medium at 4oC. The fractions were then centrifuged at 50g, the supernatant was aspirated and the pellets were pooled by resuspension in 25ml of DMEM (pH 7.4 with 0.1M NaOH) containing 2mM DTT at 4° C. The crypt suspension was then placed on a rotary inversion mixer for 20 minutes. The suspension was then spun at 50g for 2 minutes and aspirated, before final resuspension in 2ml of ice-cold DMEM. The cells were maintained on ice and used within 6 hours. The crypts were easily identified under phase contrast microscopy by their birefringence and cylindrical morphology and appeared viable as judged by their capacity to exclude 0.03% trypan blue.

Solutions

The composition of the intra- and extracellular solutions is given in table 3.1. Patch pipettes were filled with an artificial K+-rich "intracellular" solution containing 100 mg/ml of nystatin. Nystatin-containing pipette solutions were prepared freshly every 1-2 hours (from a frozen stock (10 mg/ml) of nystatin in dimethylsulphoxide (DMSO)) by adding 20m l of nystatin stock to 1ml of pipette solution in an Eppendorf tube, taking precautions to avoid exposure to light. The tube was sonicated for 2 minutes and the pipette solution was then filtered through a Dynaguard (0.2mm) ME syringe filter (Microgon inc., U.S.A.) into a foil-covered 5ml beaker kept on ice. In some experiments pipette Ca2+ was buffered to a quasi-physiological intracellular concentration [93 nM] with EGTA. The spontaneous "zero-current" reversal potential (E_m) was not affected by this manoeuvre, which can be explained by the impermeability of nystatin to divalent cations (Horn and Marty, 1988). The concentrations of Ca2+ [2 mM] and EGTA [5 mM] required were calculated according to the stability constants of Martell & Smith (1974). The tip of the pipette was filled by capillary action with nystatin-free pipette solution before the pipette was back-filled with the nystatin-containing solution. The composition of the pipette solution is as given in table 3.1.

The low Na+ solution used for extracellular ion replacement experiments was prepared by equimolar replacement of NaCl with N-methyl-D-glucamine Cl and the low Cl- solution by an equimolar replacement with Na gluconate. Changes in [K+]o were achieved by equimolar substitutions of KCl and NaCl. Hanks medium containing either 5mM TEA or 5mM Ba2+ was prepared omitting 5mM NaCl. All bath and pipette solutions were filtered through Dynaguard 0.2m m syringe filters prior to use. Osmolarities of solutions were checked using a freezing-point depression osmometer (3MO Advanced Instruments, Massachussetts, USA) and was adjusted to 295 mosmol/l by the addition of D-mannitol if necessary.

Perfusion arrangements for crypt electrophysiological recordings

A suspension of isolated crypts was aliquoted onto type 0 glass 10mm diameter coverslips (Chance Propper) precoated with 0.01% polyethyleneimine. The pre-treated coverslips were supported on 2.5cm glass micro-fibre filters (Whatman Ltd., England) and placed inside a 135mm petri dish. The crypts were left to equilibrate for 10-15 minutes at room temperature before being transferred into a purpose-built chamber containing Hanks medium, the bottom of which comprised of a type 0 glass coverslip. The chamber was mounted on the stage of an inverted microscope (Zeiss IM 35, Germany). Rapid solution changes were effected by directing a small jet of the desired solution directly onto the crypt under study, without substantially changing the composition of the bulk solution, by means of a perfusion loop similar to that described by Suzuki et al (1990). The bulk solution was constantly removed with aid of a peristaltic pump and replenished by Hanks medium flowing under gravity. All experiments were conducted at room temperature (22-25° C).

Electrical and perfusion arrangements used for crypt patch-clamp recording. The isolated crypt preparation attached to a PEI pre-coated glass coverslip is placed into a purpose-built Perspex chamber constantly perfused with Hank's medium under gravity. Excess bath fluid is removed by a peristaltic pump. The cell-perfusion device illustrated comprises a perfusion loop with a micro-puncture at the apex of the loop which is covered by a fine nozzle. This perfusion device is fed from a selected reservoir under gravity and is directed at the crypt under study using a micromanipulator. High resistance electrical seals are obtained with a polished micropipette when brought into contact with the cell membrane using a micromanipulator. The voltage-clamp circuit comprises a high gain 'feedback' amplifier (FBA) in which the output is fed back via the resistor (R) to the input, ensuring that the input voltage (V_i) is kept equal to the signal voltage (V_s). The output voltage is proportional to the current flowing through the patch of membrane in contact with the electrode.

Electrophysiological recordings

Patch-pipettes were fabricated from Boralex glass capillaries of outside diameter 1.7mm (Rochester Scientific Co., New York, U.S.A.) using a two-stage vertical pipette puller (PP-83, Narishige, Japan). The tips of pipettes were polished using a microforge before back-filling with "intracellular" solution. Patch pipettes used for current-clamp recordings had resistances of 4-5 MW when filled with KCl-rich solutions. "Zero-current" reversal potentials and whole-cell currents were recorded according to the method of Hamill et al. (1981) with a List EPC-7 patch-clamp amplifier (List Electromedical, Germany) using the perforated-patch method as previously outlined by Horn & Marty (1988). Isolated crypts attached to 10 mm diameter type 0 glass cover-slips (Chance-Propper) were mounted in a specifically designed perfusion-chamber, as described above, and the crypts were viewed on the stage of an inverted microscope (Zeiss IM 35, Germany) at a total magnification of x320.

The smooth basolateral membrane of cells in the mid-crypt region was approached with patch-pipettes without applying positive pressure to the pipette using a hydraulic micromanipulator (Narishige MO 203, Japan). Giga-ohm seals were obtained by applying light-suction to polished patch-pipettes gently pressed against the crypt membrane (see Fig.1.4.). During selected recordings, a train of small voltage-pulses (0.1-10mV) was applied to the stimulus input of the amplifier to monitor the progression of seal formation by measuring the size of the resulting current pulses. Access resistances were estimated to be in the range of 20-50 MW within 1-3 minutes of seal formation. For whole-cell current recordings, the series conductance was adjusted in parallel with the series capacitance to compensate for the whole-cell capacitive current. Access resistances were monitored and progressively compensated throughout voltage-clamp recordings. Cells were voltage-clamped at a holding potential (typically -40mV) and membrane currents were recorded in response to both depolarising and hyperpolarising voltage steps.

In current-clamp experiments, the zero-current potential of the crypt was continuously monitored to provide a measurement of membrane potential (Em). Stable Em values had been obtained, usually within 1-3 minutes of seal formation, indicating that the pipette-filling solution had equilibrated with the cytoplasm. Membrane potential changes evoked by ion substitutions and by the addition of channel inhibitors to the bathing medium were recorded, with or without the application of current pulse trains of fixed amplitude to monitor changes in membrane conductance. The zero-current reversal potentials are referred to as membrane potential values (or Em) throughout the text, assuming that junction potentials after seal formation in low access resistance perforated-patch recordings are 3/45mV (Rae & Fernandez, 1991).

The established sign convention is used throughout and potentials are reported with respect to the patch-pipette. The bath electrode consisted of an Ag-AgCl pellet. Junction potentials evoked following bath solution changes were determined and measurements obtained from recordings were corrected for the measured offsets.

Data Acquisition and Analysis

An IBM-AT microcomputer equipped with a Lab-PC laboratory interface (National Instruments, U.S.A.) and software programmes (VGEN and VCAN, J. Dempster, Dept. of Physiology and Pharmacology, University of Strathclyde, Glasgow) were used for data acquisition and the analysis of current-clamp and voltage-clamp recordings. Voltage pulse protocols and current pulse trains were applied to the stimulus input of the List EPC-7 following digital-to-analogue conversion using the Lab-PC laboratory interface. The signal from the patch-clamp amplifier was simultaneously viewed on a storage oscilloscope (Gould 1421) and recorded on videotape for subsequent analysis together with triggering pulses using an adapted pulse-code modulation encoder (Sony PCM-701ES) as described by Lamb (1985).

Replayed records of whole-cell currents were low pass filtered at 2.5kHz (-3db) using a variable 8-pole Bessel filter (Barr & Stroud) and then digitised using a PC-Lab laboratory interface. Current-clamp recordings were acquired at a frequency of 4 Hz and whole-cell recordings at 2-5kHz. Results are expressed as means ± standard errors of n observations.

APPENDIX B

Single channel recording methods for small intestinal crypts

Solutions

Carbachol was prepared as a 100mM stock in water and stored frozen. The concentrations of Ca2+ and EGTA required to buffer free Ca2+ in the bath to the desired concentration were calculated according to the stability constants of Martell & Smith (1974). The low Cl- solution was prepared by an equimolar replacement of NaCl with Na gluconate. No correction for Ca2+ chelation by gluconate was made when preparing the low Cl- Hanks medium. Hanks containing 5mM Ba2+ and 5mM TEA were prepared as for normal Hanks, but omitting 5mM NaCl. Changes in [K+]o were achieved by equimolar changes of KCl with NaCl. The composition of all other bath and pipette solutions used are given in table 8.1. All bath and pipette solutions were filtered through Dynaguard 0.2m m syringe filters prior to use.

Electrophysiological recordings

Electrophysiological recordings of single channel currents were performed as described previously with the following modifications. Smallmouth patch-pipettes used for single channel recording were fabricated from Boralex glass capillaries (Rochester Scientific Co., New York, U.S.A.) and hard typical resistances of 8-10 MW when filled with K+ or Cl- rich pipette solutions. Single channel currents were recorded as outlined by Hamill et al. (1981) using a List EPC-7 amplifier (List Electromedical, Germany) or a Biologic RK-300 amplifier (Biologic, Claix, France). Junction potentials arising from entry into the bathing solution were compensated prior to seal formation. Single channel activity was recorded from the basolateral membrane of intact crypt enterocytes once an adequate seal resistance (>10 GW) had been obtained, and the fast transient capacitive current had been compensated. To form cell-free patches the electrode was quickly retracted from the cell and the excised (inside-out) membrane patch was passed through the air-water interface to prevent vesicle formation. Cells were voltage clamped at a pre-determined holding potential and single-channel current events were recorded in response to depolarising and hyperpolarising voltage steps. In the cell-attached configuration the transmembrane potential of the patch (Vpatch) is determined by that of the cell interior (Vm) as well as that of the inside of the patch pipette (Vref) and is given by the relation:

$$V_{patch} = V_m - V_{ref}$$

Usually, Vm is unknown. The normal sign convention was observed throughout with positive (outward) currents corresponding to upward deflections of the illustrated current records.

Data Acquisition and Analysis.

Single channel currents were acquired using a voltage pulse generator programme (VGEN, J. Dempster, 1987). Stored single channel records were analysed using a semi-automated analysis programme (PAT, J. Dempster, 1987). Relay records were low pass filtered at 1.0 kHz (-3db) and digitised at 4 kHz as described in 3.2.4. Single channel currents were measured from generated unitary current amplitude distribution histograms. Channel open- and closed-state events were identified using a detection threshold crossing method with the discriminator positioned midway between the fully open and closed levels.

Channel open-state probability (Po), defined as the fraction of the total time the channel occupies the open state, was computed from the duration of channel open times determined using the detection threshold method. When patches contained more than one channel, the detection threshold was reset for each successive single channel level to determine the time intervals when 1, 2, 3,...,n channels were open simultaneously. Channel open probability for a patch containing more than one channel (nPo) was calculated as Po (level 1) + Po (level 2) ... + Po (level n). If the patch contained a known number of channel open levels the Po could be derived by dividing nPo by n.

Only membrane patches containing a single channel were used for stochastic analyses. Histograms of channel residence times were constructed from current records exceeding 40 s duration. Mean open and closed times represent arithmetic means of all observed open or shut times. Illustrated traces represent records digitised at 2-4 kHz (filtered at 0.5-1.0 kHz (-3db), Bessel 8-pole low-pass filter) using an IBM-AT computer with appropriate software (PAT, J. Dempster, 1987) and then imported as ASCII files into a graphics programme (Sigmaplot 5.0, Jandel Corporation, 1992).

Current-voltage relationships were analysed assuming that they obeyed the Goldman-Hodgkin-Katz constant field theory (GHK), which is a good approximation of the behaviour of many channels not exhibiting anomalous rectification. The data were fitted to equations of the form:

$$i = g.V \frac{\left[C_o - C_i.\exp\left(\frac{FV}{RT}\right)\right]}{\left\{C_{sym}\left[1 - \exp\left(\frac{FV}{RT}\right)\right]\right\}}$$

where (i) is the single channel current, G is the single-channel conductance, C_o and C_i are, respectively, the ion concentrations bathing the extracellular and intracellular sides of the membrane patch, V is the command potential in the case of excised patches or the sum of the command potential and membrane potential in the case of cell-attached patches, R is the gas constant, T is the absolute temperature, and F is Faraday's constant. Conductances measured in the presence of asymmetrical ion distributions are normalised to the symmetrical case by dividing by C_{sym}.

APPENDIX C

Limitations of the voltage clamp technique

Much of the early modelling of classical electrophysiology, developed using the squid giant axon, was based upon the physics of undersea cables which were conductive cores sheathed by an insulator (*which serves as an excellent analogue of a nerve axon*) which conduct voltages over long distances and face the problems of signal and voltage loss. This led to the development of the cable equation and explained how the nodes of Ranvier serve as electrical substations, boosting the attenuated signal at periodic intervals along the length of the axon. It should, however, be noted that many cells have evolved to be compartmentalized, notably neurons, wherein they are not truly isopotential across all parts of the cells, which effectively enables sub-computation and integration of inputs to occur at key 'decision' points such as the axon hillock which initiates action potentials and at the synapse, where the non-linear summation of a neuron's various inputs occurs.

The creation of an effective 'space clamp' serves to eliminate voltage gradients along the axon which will invariably distort the amplitude of the currents that are measured. Factors contributing to the limitations of an effective space clamp are as follows:

[1] Series resistance error (Rs)
As we have discussed before, the series resistance (Rs) is that resistance which is encountered by the electrical circuit of the amplifier when it attempts to impose a voltage step upon the cell, thereby effectively charging the membrane capacitance of what is, for all intents and purposes an RC circuit. An instantaneous change in membrane potential requires an almost infinite current which is, of course, practically impossible.

The series resistance effectively slows the capacitive currents thereby limiting the rate at which the membrane capacitance can be charged. These capacitive currents may overlap with the ionic currents of the whole cell membrane, effectively distorting their measurement.

Although the series resistance issue is practically overcome by the patch clamp technique, given that it introduces a GΩ seal and a relatively low value of Rs relative to so-called 'sharp electrodes', there are, nonetheless, issues arising when the distance or the capacitive area of the cell (or electrically coupled cells) is taken into consideration. Large cells with a low input resistance are effectively impossible to voltage clamp in their entirety, although we can passively measure the currents and action potentials that flow through them.

This brings us on to the issue of the space clamp and reintroduces the all-important concept of the space, or length constant λ.

[2] The space, or length constant, λ

As previously discussed, the space, or length constant λ, is defined as the distance over which the steady-state voltage decays to $1/e$ (i.e. 37 %) of its original value in a hypothetical cable. The value of λ, a quantification of steady-state voltage decay as a function of distance within a cell, is given by the following equation:

$$\lambda = (R_m \cdot d/(4R_a))^{1/2}$$

Thus, if the value of λ is substantial and there is a small series resistance error, a substantial input resistance (R_m) and a comparatively low axial resistance (R_a), then we may record with some confidence. However, if the value of λ is small, then we are recording in vain.

APPENDIX D

Expression of GABA$_A$ receptor subunits in PC12 cells

GABA$_A$ receptors: structure and function

Members of a superfamily of ligand-gated ion channels including nicotinic ACh, 5-HT$_3$, GABA$_A$, GABA$_C$ and glycine receptors, which likely evolved from a common ancestor. GABA$_A$ receptors are pluripotent drug targets assembled from a repertoire of more than 18 subunits including α1-6, β1-4, γ1-4, δ & θ.

Binding of GABA to GABA$_A$ class receptors increases membrane permeability to chloride ions usually resulting in membrane hyperpolarization and a reduction in excitability. The stoichiometry of the GABA$_A$ receptor is pentameric, with 2α + 2β subunits forming a Cl-selective pore in combination with either a γ or δ subunit
GABA$_A$ receptors are present in adrenal chromaffin cells, but the subunit composition was previously unknown.

Conventional RT-PCR was thus used to determine which GABAA subunits are expressed in the pheochromocytoma cell and adrenal gland. It was determined that growth factors such as NGF, PACAP-38 and insulin induce or modulate the abundance of GABAA receptor subunits in the PC12 cell and that the GABA$_A$ α and γ subunit expression profile in the PC12 cell line was essentially consistent with that in the adrenal gland. Moreover, the subunit expression profile is consistent with that believed necessary for the formation of functional receptor-channels, and the presence of colocalized GABAA receptor subunits upon the surface of the PC12 cell was confirmed by immunocytochemistry. These findings demonstrate the utility of the PC12 cell as a model for the regulation of gene expression and GABAergic function within the adrenal gland.

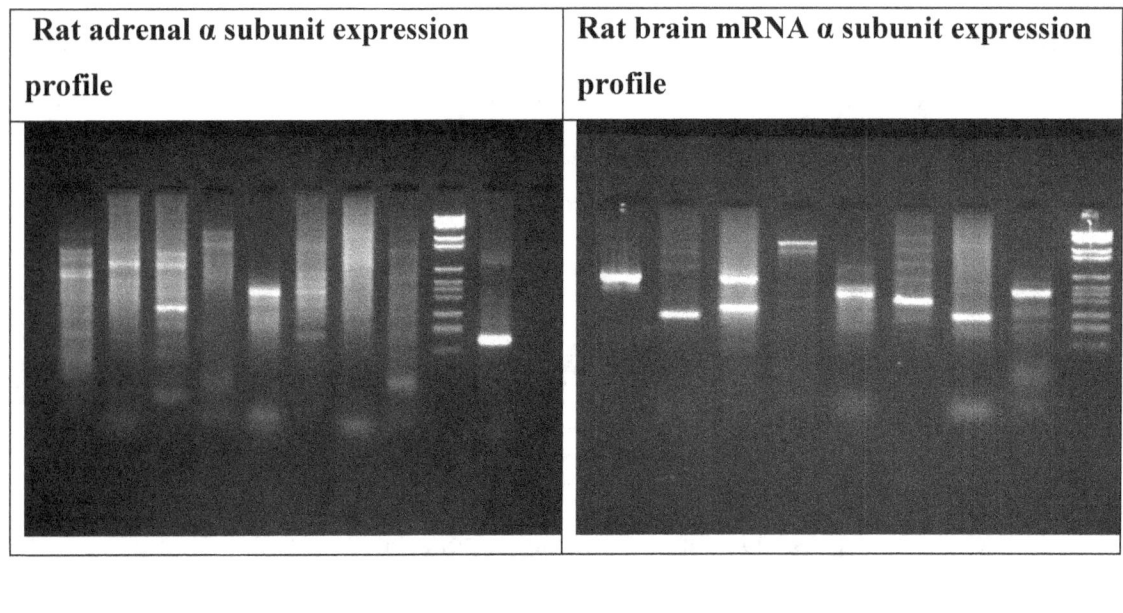

subunit	α1	α2	α3i	α3ii	α4i	α4ii	α5	α6
exons	E1-6	E1-4	E1-6	E6-8	E6-8	E1-4	E1-4	E1-4
size	580	345	601	484	503	387	292	476

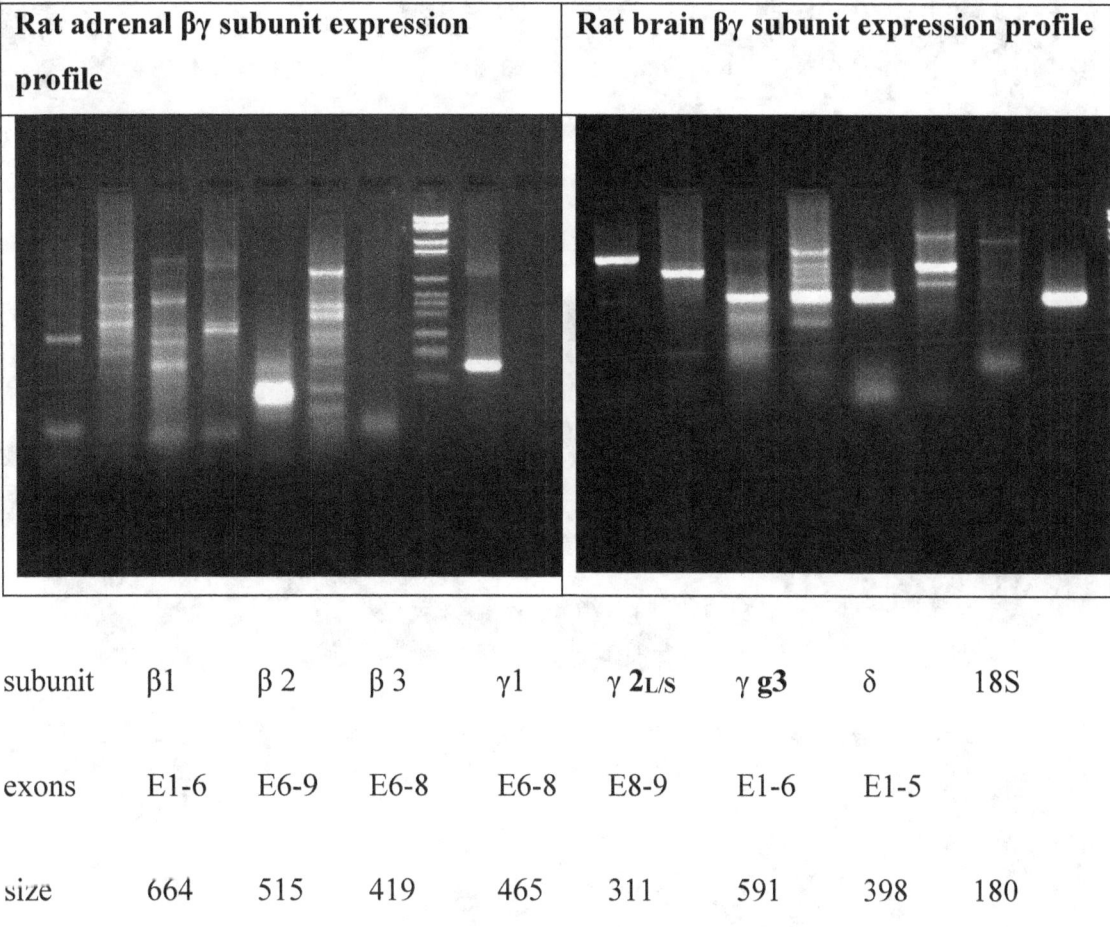

subunit	β1	β2	β3	γ1	γ 2L/S	γ g3	δ	18S
exons	E1-6	E6-9	E6-8	E6-8	E8-9	E1-6	E1-5	
size	664	515	419	465	311	591	398	180

My analysis indicated that the adrenal medulla appears to express α3, α4, γ2 and γ3 subunits, but, surprisingly, no β subunits could be detected. An equivalent analysis for the PC12 cell stimulated by a range of growth factors indicated that the PC12 cell, like the adrenal, expresses α3 and α4 subunits, and that their expressed may be regulated by growth factors (*see below*). However, in contrast to the findings for the adrenal, the PC12 cell appears to express both β2 and β3 subunits (not shown) which are expressed on the surface of the cell (*see below*).

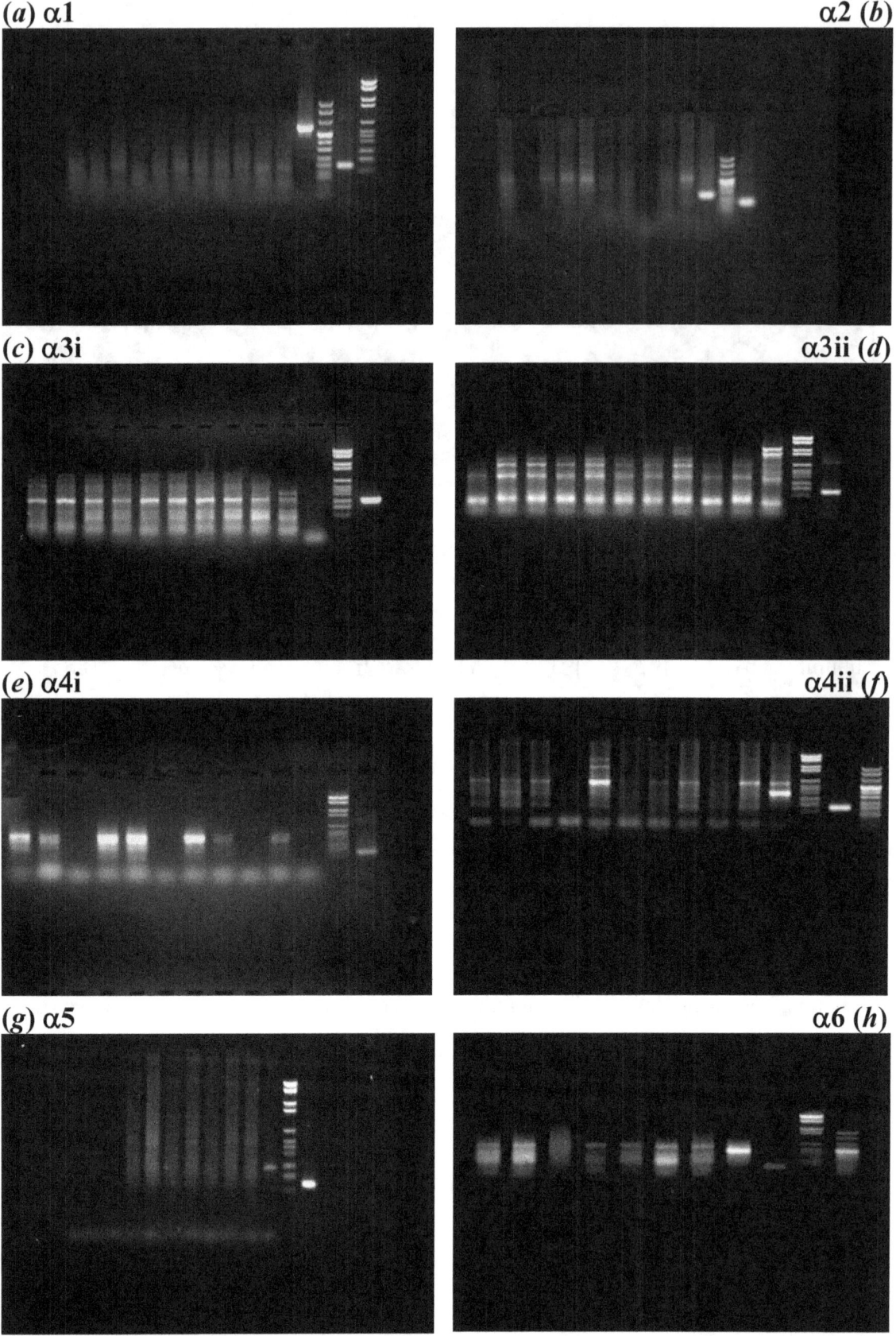

Expression of α subunits in PC12 cells above and β & γ subunits below in response to a panel of growth factors.

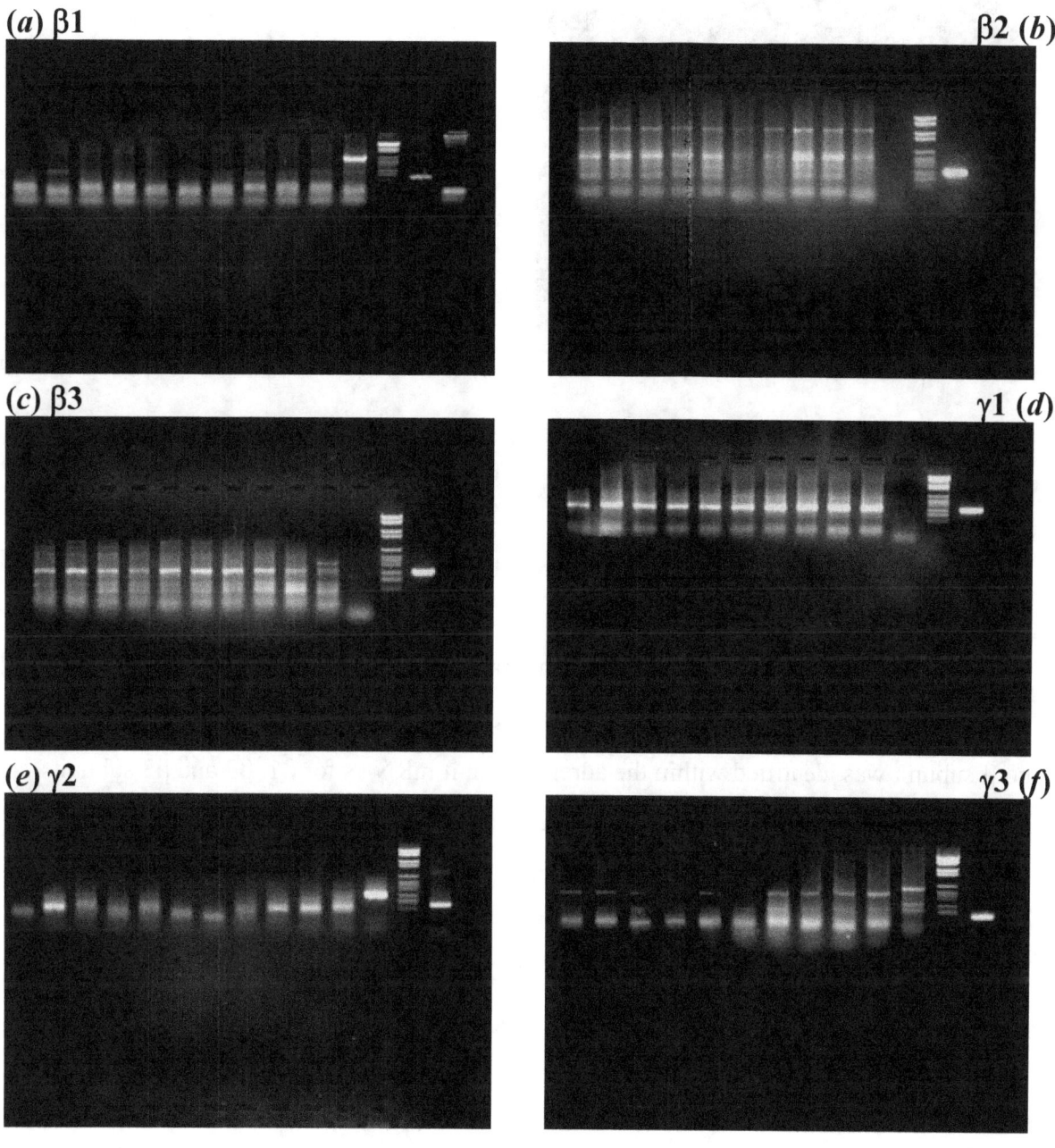

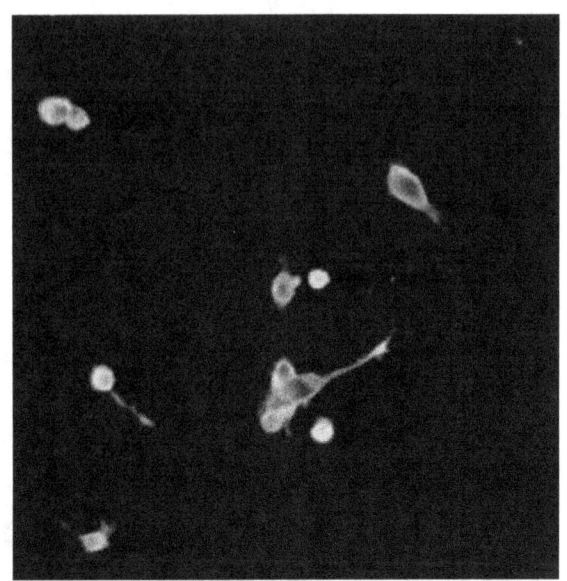 Immunocytochemistry demonstrates β subunit labelling in PC12 cells.

Summary

$GABA_A$ receptors play a pivotal role in the regulation of excitability and secretion. Although it was known that adrenal chromaffin cells possessed functional $GABA_A$ receptors, this was their first classification in terms of subunit composition. Both serum-deprived PC12 cells and the rat adrenal express α3, α4, γ1, γ2 and γ3 subunits although no β subunit was identified within the adrenal even if mRNAs for β1, β2 and β3 subunits were expressed in the PC12 cell line which also derives from the rat. Immunocytochemistry reveals the colocalized surface expression of both β and γ subunits within the PC12 cell and insulin, Nerve Growth Factor (NGF) and Pituitary Adenylate Cyclase Activating Polypeptide (PACAP-38) appear to modulate the level of expression of α, β and γ subunits within PC12 cells.

APPENDIX E

Data for excised patch ion substitution experiments

Composition of solutions used in ion substitution experiments

	Hanks	0.1 K$^+$	1 K$^+$	0.1 K$^+$	10K$^+$/135Na$^+$	145KCl/low Ca^{2+}	145 KCl/high Ca^{2+}	10mM K Na$^+$ free
NaCl	140	-	-	145	135	-	-	-
KCl	5	-	-	0.1	10	145	145	10
CaCl$_2$	1.3	1.3	1.3	0.1	0.1	0.1	1.91	0.1
EGTA	-	-	-	0.25	0.25	0.25	2	0.25
MgCl$_2$	0.5	0.5	0.5	0.5	0.5	0.5	0.5	0.5
HEPES	10	10	10	10	10	10	10	10
Na Gluconate	-	145	145	-	-	-	-	-
K Gluconate	-	0.1	1	-	-	-	-	-
K$_2$HPO$_4$	0.36	-	-	-	-	-	-	-
KH$_2$PO$_4$	0.44	-	-	-	-	-	-	-
NaHCO$_3$	4.2	-	-	-	-	-	-	-
N-methyl-D-glucamine Cl	-	-	-	-	-	-	-	135

All concentrations are given in mM and solutions titrated to pH 7.2 with Tris buffer.

Table 2

Change in composition of intracellular solution (concentrations are given in mM, $\{[Ca^{2+}]_i\}$)	Mean change in I_{out} (pA)	± SEM	n
10 KCl, 135 NaCl {36nM} ® 145 KCl {36nM}	2.67	0.54	7
10 KCl, 135 NaCl {36nM} ® 145 KCl {1m M}	1.71	0.26	4
10 KCl, 135 NaCl {36nM} ® 10 KCl, 135 NMDGCl {36nM}	-5.47	1.13	6

Mean changes in outward current (I_{out}) given in pA (± SE) for n different excised inside-out basolateral patches obtained with pipettes of resistance 8-10 MΩ containing a solution of composition (mM) {1 KGluconate, 145 NaGluconate, 1 CaCl$_2$, 0.5 MgCl$_2$, 10 HEPES, pH 7.2 with Tris base}. Changes in the values of outward current are presented as means of paired values (Change in I_{out} = I_{out} {test solution} - I_{out} {10 KCl, 135 NaCl, 36nM [Ca^{2+}]$_i$ control}).

Table 3

Change in composition of intracellular bathing solution (concentrations are given in mM, $\{[Ca^{2+}]_i\}$)	Mean shift in reversal potential from 10KCl, 135NaCl control (mV)	± SEM	Junction potential (mV)	n
10 KCl, 135 NaCl, {36nM} ® 145 KCl, {36nM}	-10.0	1.4	-0.1	4
10 KCl, 135 NaCl, {36nM} ® 145 KCl, {1m M}	-10.8	0.95	-0.1	4
10 KCl, 135 NaCl, {36nM} ® 10 KCl, 135 NMGCl, {36nM}	+31.0	4.2	-1.5	4

Mean changes in the extrapolated zero-current reversal potentials upon the solution changes indicated, given in mV (± SE), for n different excised inside-out patches with 8-

10 MΩ pipettes filled with solution of composition (mM) {1mM KGluconate, 145mM NaGluconate, 1 CaCl$_2$, 0.5 MgCl$_2$, 10 HEPES, pH 7.2 with Tris base}. Shifts in reversal potentials are presented as means of paired values of the test solution compared to the 10 KCl, 135 NaCl control solution (see also Fig.3.B.). Junction potentials measured in the voltage-clamp mode are given.

APPENDIX F

Experimental methods for recording from β-cell clusters

Cell culture

Islets of Langerhans were isolated using a modification of the methodology of Rorsman et al., 1991. Two to five fed female NMRI mice (25-30g) were injected with 0.7-1ml pilocarpine and after 45min-1 hr incubation mice were cervically dislocated, decapitated and briefly immersed in 70% ethanol. The pancreas was surgically exposed and injected with a sterile-filtered Rorsman isolation buffer (from frozen) containing (mM); NaCl 135, KOH 4.8, $CaCl_2$ 2.5, $MgCl_2$ 1.2, HEPES 10, glucose 10, bovine serum albumin (BSA) 1% (spontaneous pH 7.29), dissected and then incubated briefly in 10ml of isolation buffer supplemented with 50μg/ml Fungizone (Gibco). Pancreatic tissue was then finely chopped and digested with 5-7mg collagenase, in proportion to tissue yield, and incubated for 12 min with gentle shaking in a water bath at 37°C. The resulting digest was washed first with DNAse-containing isolation buffer at 4°C to quench the reaction, and then washed twice with chilled Rorsman isolation buffer. Islets (typically 100-200) were sorted from the exocrine tissue and transferred to and incubated for eight min in a Ca^{2+}-free saline containing BSA and EGTA at 4°C. The islets were then centrifuged at <100g for 5 min before decanting the supernatant, which was replaced with 2ml RPMI 1640 medium (Gibco BRL, Paisley, Scotland, U.K.) supplemented with 10mM glucose, 100 IU/ml penicillin, 100μg/ml streptomycin, 10% heat-inactivated foetal bovine serum and L-glutamine. The islets were then triturated with a heat-polished, sialinised pasteur pipette until only a fine cell suspension remained. The resulting suspension was plated at 0.3 ml/coverslip and allowed to attach for 7-20 hours before supplementation with 3ml RPMI 1640 medium.

β-cells comprise approximately 80% of islet cells in normal mice (Hedeskov, 1980) and have a larger diameter (11-14μm) than β-cells (8-11μm, Pipeleers, 1987) and were thus co-cultured as a heterogeneous population of single-cells and small clusters. Recordings were routinely made two days after trituration, except in recordings from identified single-cells, which were performed using an acutely isolated preparation wherein the RPMI trituration medium was additionally supplemented with 5mM HEPES and the pH adjusted to 7.4 with 1N NaOH in order to prevent cellular acidosis during isolation. Cells from selected coverslips were then transferred directly into an appropriate Krebs recording media after pre-incubation for 40-60 min in low glucose (G3) Krebs medium within a 37°C CO_2 incubator. Cells at the centre of the clusters were selected to optimize the space-clamp.

Solutions

All cells and clusters were perfused in a heated chamber prior to recording with Krebs ringer containing (mM) NaCl 120, KCl 4.8, $CaCl_2$ 2.5, $MgCl_2$ 1.2, $NaHCO_3^-$ 24, HEPES 5 (pH 7.35 with 1M NaOH) after gassing for 15 min with 95%O_2 + 5%CO_2, with a D-glucose concentration appropriate to the experiment. Solutions were continuously gassed during the recording with 94%O_2 + 6% CO_2. The K150 pipette solution contained (mM) KOH 140, KCl 10, N-methyl-D-glucamine 60, $MgCl_2$ 1.2, HEPES 10 (pH 7.1 with 1M H_2SO_4), and the Cs150 pipette solution contained (mM) CsOH 140, CsCl 10, N-methyl-D-glucamine 60, $MgCl_2$ 1.2, HEPES 10 (pH 7.1 with 1M H_2SO_4).

Ion substitution experiments were performed by equimolar substitutions of constituents within the Krebs medium. The low (59mM) Na^+ Krebs contained (mM) NaCl 35, KCl 4.8, $CaCl_2$ 2.5, $MgCl_2$ 1.2, $NaHCO_3^-$ 24, N-methyl-D-glucamine 85, HEPES 5 (pH 7.35 with 1M NaOH), and the low (0.25mM) Ca^{2+} Krebs contained (mM) NaCl 120, KCl 4.8, $CaCl_2$ 0.25, $MgCl_2$ 3.45, $NaHCO_3^-$ 24, HEPES 5 (pH 7.35 with 1M NaOH).

Experiments performed to replicate and explain the experiments of Smith et al., (1989) were identical to those published, the 'Smith external' buffer containing (mM) NaCl 138, KCl 5.6, $CaCl_2$ 2.6, $MgCl_2$ 1.2, TEACl 10, HEPES 10 (pH 7.4 with 1M NaOH) and the TEA-free buffer contained (mM) NaCl 138, KCl 5.6, $CaCl_2$ 2.6, $MgCl_2$ 1.2, HEPES 10 (pH 7.4 with 1M NaOH). The Smith hypotonic pipette solution contained (mM) KCl 10, CsCl 10, Cs_2SO_4 70, $MgCl_2$ 7, HEPES 10, pH 7.4 with CsOH. Measurements from the air-stone gassed reservoirs confirmed that the pH of the solution was maintained in the range of 7.35-7.4 by continuous gassing with the 94%O_2:6% CO_2 mixture.

Voltage-clamp recordings

Recordings were made with an EPC-9 patch-clamp amplifier using HEKA v8.11 voltage-clamp software (HEKA Electronics, Lambrecht/Pfalz, Germany). Cells plated upon coverslips were maintained in a purpose-designed Perspex chamber equilibrated at 30-33°C with a heated stage (Intracell, Royston, Herts.). The temperature was confirmed by measurement with a thermistor in the bath and at the mouth of the perfusion pipette before and after each recording. Clusters were viewed under phase-contrast optics at 400x on an Axiovert 100 microscope, and for single-cell recordings using non-phase optics and an oil-immersion objective (x400).

Perforated-patch recordings were performed after the original method of Horn and Marty (1988) using WPI glass electrodes with a 2-4 MΩ resistance in saline after fire-polishing. After coating with Sylgard elastomer (Dow Corning, Wiesbaden, Germany) to reduce stray capacitance and filling the tip with antibiotic-free intracellular solution, pipettes were back-filled with K150 or Cs150 solution containing Amphotericin B. The principal recording solution was a HCO_3^--buffered Krebs solution supplemented with 5mM HEPES which was adjusted, after gassing for 10-15 min with 94% O_2 + 6% CO_2, to pH 7.35 with 1N NaOH. All Amphotericin B containing pipette solutions were made freshly every hour with 250µg/ml Amphotericin B (Sigma), from a fresh stock of Amphotericin B dissolved by sonication in Hybrimax DMSO (Sigma). Fresh Amphotericin B-containing pipette solutions were sonicated for 5min in 0.5ml pipette solution in an Eppendorf and were maintained wrapped in foil in a syringe applicator on ice. All recordings were performed in the dark with only a minimum illumination intensity allowed during seal formation and perfusion alignment, due to the known light sensitivity of the polyene antibiotics.

The fundamental voltage-clamp pulse protocol used throughout these experiments, after the original methods of Hamill et al., 1981, was termed IK1, and comprised a series of 21 voltage steps applied in 3mV increments from –70 to –10mV, each 300ms in duration from a holding potential of -70mV with a 2s inter-pulse interval. All other protocols were as described in the figure legends. The command voltage was checked weekly using an oscilloscope in parallel to the command output of the amplifier. All recordings were commenced 10-15 min after seal formation at the central point of the small cluster to allow permeabilization of the patch and equilibration between the intracellular milieu and the pipette solution. Recordings were monitored continuously until both currents in G3 and the access conductance had stabilized at a minimum (typically >50nS). All recordings were acquired at a frequency of 10 kHz and presented after low-pass filtering at 1-2 kHz by an internal 8-pole Bessel filter.

In all recordings (except those replicating the experiments of Smith et al.1989) clusters and cells were perfused directly at 0.7-1.0 ml/min with solutions passed through a heated block to equilibrate at 30-33°C into an estimated chamber volume of 0.8-1.2ml constantly perfused with Krebs HEPES/HCO_3^--buffered medium gassed continuously with 94%O_2:6%CO_2. All solutions contained 24mM $NaHCO_3$ and 5mM HEPES with a measured pH in the range 7.35-7.4, kept constant by continuously gassing all reservoirs with 94%O_2:6%CO_2 using miniature air stones. Solutions were changed using an electronic valve-regulated (24V Isolatch valves, General Valve Corporation, N.J.) perfusion system directed via a heating block into a fused silica capillary delivery flow pipe (O.D.740µm, Composite Metal Services Ltd., Worcs, U.K.,), ensuring that rapid and direct solution changes were made with only a minimal intermixing of solutions. The estimated change time of the solution at the nozzle was less than 1s with a dead time of 20s. Both the bath temperature and the temperature of the outflow of the perfusion nozzle were measured before and after each recording using a micro-thermistor, and the temperature was adjusted so that it was between 30 and 33°C, although a variation of 1-2°C between the values taken at the beginning and the end of the recording and of 1°C between the applicator nozzle (lower) and the bath were typical, and hence absolute values are not given for each recording. The mean temperature recorded at the nozzle was 31.4°C ± 0.8°C (n=30).

Analysis

Equilibrium potentials for Ca^{2+} were calculated according to Nernst in the form:

$$ECa = \frac{RT}{2F} \cdot \ln\left(\frac{[Ca]o}{[Ca]i}\right)$$

Where E_{Ca} is the equilibrium potential for Ca, R is the general gas constant (8.315 J.K^{-1}mol^{-1}), T is the absolute temperature (K), F is Faraday's constant (9.648 x 10^4 C.mol^{-1}), [Ca]$_o$ is the extracellular Ca^{2+} concentration and [Ca]$_i$ is the intracellular Ca^{2+} activity.

Fits of repolarization potential to paired-pulse current amplitude ratios were to the Boltzmann equation in the form,

$$\frac{P2}{P1} = \left[\frac{\frac{P2}{P1}min - \frac{P2}{P1}max}{1} + e\left(V - \frac{V0.5}{dV}\right)\right]$$

where P_2/P_1 is the ratio of the second peak current amplitude elicited by a depolarizing step (P$_2$) divided by the initial peak current amplitude from baseline (P$_1$); P$_2$/P$_1$ $_{min}$ is the minimum ratio elicited with no recovery pulse (0), P$_2$/P$_1$ $_{max}$ is the maximal recovery amplitude elicited with a repolarization step to –80mV, V is the repolarization step potential, V0.5 is the repolarization potential at which a 50% recovery of current is elicited and dV is the change in repolarization potential.

Exponentials were to capacitive current decays were fitted to the form:

$$IC = Imin + Io.exp - \left(t - \frac{to}{\tau}\right)$$

where I$_C$ is the capacitive current at time t, I$_{min}$ is the whole-cell current to which the transient decays, I$_o$ is the initial capacitive current at t$_o$ (zero time) and τ is the time constant for a first order decay, given by R$_s$.C$_m$, where R$_s$ is the access resistance and C$_m$ is the membrane capacitance.

Current clamp recordings

Zero-current reversal potentials (Em) were measured in the current-clamp recording mode using the Amphotericin B perforated-patch recording technique with patch electrodes either containing the K150 solution or one containing 130mM K^+ with NMG^+ as the balancing cation and osmolyte. K150 was estimated to be the intracellular K^+ activity from zero-current reversal potential measured in G3 (3mM glucose), conditions under which K^+ conductance dominates the resting membrane potential, and then Em values were determined both when the external K^+ concentration was kept at 4.8 mM (K4.8) and increased to 60 mM (K60). Em values measured before and after K^+ substitution were used to estimate the intracellular K^+ activity according to Nernst, assuming that the β-cell membrane is perfectly K^+-selective in G3.

Calcium imaging

Cultures were pre-incubated for 60 min with 1μM Fura-2-AM in G3 Krebs at 37°C supplemented with 1% BSA. Cells were washed with Krebs and viewed with a 40x oil-immersion objective and the emission after excitation at 340/380 using a monochromator was measured at 510nm after passage through a band-pass filter. The system and software used were supplied by PTI Systems.

Single-cell recording and selection

Trituration of islets gave rise to a culture of variable composition. However, what appeared to have been large single cells with a perfectly spherical and symmetrical appearance under 400x phase-contrast optics were seen under an oil-immersion objective to be in fact small clusters containing 2 to 7 cells with symmetrical apposing membranes and clearly-defined nuclei. Observations from acutely isolated preparations indicated that, for the first 5-8 hours of culture, individual cells retain their morphology and small clusters of cells at first resemble chains of bacterial cocci. They then progressively attach to the cover-slip and merge to form apparently perfectly spherical islands within 8-12 hours after trituration. Caution should thus be taken to distinguish single-cells from such small clusters after one to two days in culture by avoiding phase-contrast or Hoffmanized optics.

Isolated single-cells were thus identified under oil-immersion optics at 400x with a second investigator verifying 'single-cell status' after each recording. To avoid any dedifferentiation of single-cells in prolonged primary culture, single-cells were acutely isolated by a more extensive trituration in an RMPI 1640 medium supplemented with 5mM HEPES (pH 7.4) to avoid any changes in gene expression resulting from cellular acidosis during isolation. All cells were recorded from within 2-8 hours of plating, the point at which strings of single cells began to merge and form clusters. The recording conditions were otherwise identical to those used for clusters.

Limitations imposed by inability to perform capacitance measurements

Using the automated capacitance compensation feature of the HEKA software and the EPC-9 amplifier system (which lacks analogue dials), C_{fast} was first neutralized after introducing the electrode into the bath, after which C_{slow} was compensated electronically within 3-5 min of seal formation. By this method inaccurate parameter values of <3pF and <2 MΩ for C_{slow} and R_{series} respectively were typically obtained. This was due to the inability of the HEKA system to accurately fit Cm and Rs values for small cells (Dr. Francisco Mendez, HEKA, personal communication). Overcompensation of either R_{series} or C_{slow} parameters normally results in a 'ringing', or oscillation of the electrode with a concomitant loss of the membrane seal. The automated parameters appeared to effectively compensate the capacitive transient without causing electrode 'ringing', despite the implausibility of the series resistance and C_{slow} values reported. However, due to the small capacitance of β-cells ($\cong$ 6pF, Mariot et al., 1998), the low access resistances obtained and the high input resistance of the β-cells in G1/G3 (2-4 GΩ), the resulting small, fast transient exceeded the lower limit of resolution of the HEKA exponential fitting program, using the relationship:

$$\tau = Rs.Cm$$

However, the predicted upper limit of uncompensated series resistance errors will be less than 1mV if the estimated upper limit from the fitting of transposed traces into Microcal Origin for Rs is taken as 30MΩ and the lower limit measured for R_m was 2GΩ.

APPENDIX G

Recording GABAergic currents from Xenopus *oocytes*

Methods

Oocyte preparation

Xenopus laevis (Xenopus I, Ann Arbor, Michigan, USA) were anesthetized by hypothermia and oocytes were surgically removed from the frog and placed in Oocyte Ringer (OR$_2$) which comprised (mM) 82.5 NaCl, 2.5 KCl, 10 HEPES, 1 CaCl$_2$, 1 MgCl$_2$, 1 Na$_2$HPO$_4$, 50 U/ml penicillin and 50µg/ml streptomycin, pH 7.5 with 2N NaOH. Oocytes were dispersed by dissection and incubated in Ca^{2+}-free OR$_2$ containing 0.3% collagenase A (Boehringer Mannheim, Indianapolis, IN) for 75 min. After rinsing thoroughly with OR$_2$, stage VI oocytes were manually sorted and injected with cRNAs before storing overnight at 18°C.

Site-directed mutagenesis

Oligonucleotide-directed mutagenesis was conducted according to the manufacturer's protocol (GeneEditor *in vitro* site-directed mutagenesis system, Promega Corporation). *In vitro* transcription was achieved from an SspI-linearized template using the T7 MegaScript transcription kit (Ambion, Austin, TX) and the relative concentrations and qualities of cRNA estimated on a denaturating formaldehyde-containing agarose gel, so that equivalent amounts of RNA could be injected. Successful mutagenesis was verified by DNA sequencing after linearizing the DNA with SspI and by verifying changes in the engineered restriction endonuclease digest pattern upon mutagenesis.

Electrophysiology

After 3 days incubation at 18°C currents evoked by application of agonist were recorded from oocytes by the standard two electrode voltage-clamp method (**npi** Turbo TEC-05 TEVC amplifier, Adams and List, Westbury, NY) at a holding potential of −70mV. Data were acquired at 20-100Hz using an ITC-16 interface with HEKA v8.30 software (Heka elektronik GmbH, Lambrecht, Germany). Recording electrodes were pulled from a filament-containing electrode glass of diameter 1.0mm (A-M Systems Inc, Carlsborg, WA, USA). The passive (membrane potential) recording electrode typically was of 0.7-1.0 MΩ resistance in OR$_2$ after filling with 3M KCl, whilst the active (current injecting) electrode was of 1.2-1.8 MΩ resistance, with a combined resistance of no greater than 2-2.5 MΩ. This improved the longevity of recording through reduced oocyte dialysis without leading to electrode instability and oscillations associated with higher input resistances.

Perfusion systems

To obtain concentration-response relations agonist was applied to a small volume chamber (65μl) using a continuous perfusion manifold fed under gravity from up to 20 glass chambers selected by five-way valves feeding into a four-way valve with a common outlet port. Measurement of perfusion kinetics with phenol red (at 630nm) using a high-speed fraction collector and a plate reader gave a measured $t_{1/2}$ of 3s and a clearance time of less than 10s (upper estimates due to mixing in collection tubing).

To address the issue of multiple exponential fittings being attributable to a perfusion artifact, a fast perfusion method was realized by the direct parallel port (modified catheters) application of solutions into the 60μl chamber. This was achieved by means of two 14V solenoid valves arranged electronically in parallel and driven by a Low Noise Power Supply to allow instantaneous switching between agonist and agonist-free solutions. With a measured chamber flow rate of 195μl/s and a chamber volume of 60 ± 5μl the $t_{1/2}$ was reduced to 0.31s, with a 99% solution exchange time of approximately 2 s assuming complete chamber equilibration.

Analysis

For dose-response relations current amplitudes were fitted to the normalized form of the Hill equation using Microcal Origin v.6.0 software in the form;

$$I = I_{max} / (1+[EC_{50}/\{A\}]^n)$$

Where **I** is the peak current evoked after application of agonist at concentration {A}; I_{max} is the peak current evoked at saturating agonist concentrations, the EC_{50} is the agonist concentration at which a half-maximal current response is evoked and **n** is the Hill coefficient. Data were fit to the Hill equation after first normalizing the data to the extrapolated maximum value (I_{max}), the means of the data are presented ± SE as a function of agonist concentration for the number of experiments given.

The descriptive kinetic analysis took the form of measuring the time taken to achieve a half-maximal current deflection from the current baseline prior to application of the agonist ($t_{1/2}$ of activation) or closure from full steady-state activation (equilibrated agonist-receptor state; $t_{1/2}$ of deactivation) and were determined using the HEKA v8.30 software. Half-times for channel activation and deactivation were plotted as a function of the effective agonist concentration determined by dividing each agonist concentration tested by the EC_{50} derived from the same series of agonist-current response amplitude relationships.

Exponential fitting

Exponential fits to current deactivations were derived by fitting to the equation of the general form given below for one, two or three exponentials within a region delimited by a manually positioned cursor using the HEKA v8.3 software. For the exponential fits to the current traces shown, the values obtained were also verified by importing and fitting in Origin v6.0 (Microcal Software Inc., Northampton, MA, USA). Equations for fitting to three exponentials (current deactivation) in HEKA pulsefit were of the form:

$$I(t) = a_0 + a_1\exp(-t/\tau_1) + a_2\exp(-t/\tau_2) + a_3\exp(-t/\tau_3)$$

where a_0 is the common starting amplitude of the exponential or amplitude to which the exponential decays (1-exponential), $a_{1...3}$ and $\tau_{1...3}$ are the individual starting amplitudes and the time constants of the 1st, 2nd and 3rd exponentials respectively. Differing starting amplitudes imply that exponential decays were sequential, i.e. dependent upon a progressive transition from one open state to another. For fits to activations (1-exponentials) equations of the general form shown below were used:

$$I(t) = a_0 + a_1(1-\exp(-t/\tau_1)) + a_2(1-\exp(-t/\tau_2)) + a_3(1-\exp(-t/\tau_3))$$

Determination of the quality of fits (residual) analysis was done either in HEKA PulseFit v.8.30 according to a Simplex Optimization algorithm (M. S. Caceci & W. P. Cacheris, 1984, Byte, 340 ff) where the current 'residual' was calculated from the expression:

$$\sum\{\sqrt{[(I_M - I_F)^2]}\}$$

where IF is the value determined by the exponential fit whereas IM is the actual or measured value derived from the digitized recorded current decay. Values are thus given in pA, or otherwise residuals were determined following export of HEKA ascii current files to Microcal Origin 6.0 where unit-less residuals were determined after fitting with a non-linear least squares fitting algorithm. Residuals were determined for fits to one, two and three exponentials with starting amplitudes (a1…3) for each exponential (τ1…3) set as a free variable. Differing starting amplitudes suggest sequential channel state transitions.

Residuals provide a test of the validity of modeling a fit of receptor-channel deactivation to a single or multiple exponentials.

APPENDIX H

Studying ion channels through cell volume regulation

Volume regulation in epithelia and endothelia

During transepithelial solute flow epithelial cells maintain their volume in the face of a virtual torrential flood, and so must constantly readjust the transport processes mediating solute entry into and exit from the cell, and the measurement of cell volume functions as a sensitive indicator of the balance of these two processes.

Regulation of cell volume in endothelium and epithelium is critical in the regulation of fluid and electrolyte transport between tissue compartments (e.g. intestinal lumen and villus microcirculation; cerebrospinal fluid and cranial microvasculature). In the vascular endothelium, factors that cause endothelial shrinkage (e.g. hypertonicity) or retraction (e.g. histamine) are known to increase endothelial permeability. Although large step changes in extracellular osmolarity may be uncharacteristic of the physiological and pathophysiological states that non-intestinal cells are exposed to; in the intestine it is likely that the epithelium will routinely encounter large fluctuations in luminal osmolarity and will, therefore, require cell mechanisms to recover changes in cell volume rapidly, as intestinal epithelial cells must coordinate not only ion fluxes that regulate volume, but also those responsible for transcellular solute transport. In other cell types, studies of volume regulation have been performed more "physiologically" by exposing cells to gradual changes in extracellular osmolarity wherein they undergo isovolumetric regulation (IVR) up to a threshold osmolarity beyond which point they shrink (Lowr et al., 1989).

Regulation of volume in response to hyperosmotic media

Cell shrinkage in response to a hyperosmotic challenge stimulates net ion and osmotically obliged water uptake in many cell types leading to a restoration of "normal" cell volume, a process known as regulatory volume increase (RVI). Cells capable of RVI in response to RVI can be distinguished into 2 categories, those that can undergo RVI in response to a direct elevation in extracellular osmolarity (e.g. *Necturus* gallbladder cells) and those that only undergo RVI only in response to the RVD/RVI protocol; in which the cell is exposed first to a hypotonic and then to an isotonic medium (e.g. mouse pancreatic ß-cells, Ehrlich ascite tumour cells and intestinal 407 cells). Such cell types appear incapable of RVI in response to direct hypertonic challenge. In vitro in invertebrate cells nonelectrolytes play a major role in the RVI response; whereas in vertebrate cells the RVI response is accounted for predominantly by electrolytes.

RVI is generally reported to be mediated by stimulation of either Na-K-2Cl cotransport, NaCl cotransport or Na^+-H^+ exchange (cite review). Na-K-2Cl cotransport systems have been defined by 3 criteria:(1) interdependency of K^+, Na^+ and Cl^- ions present on the same side of the membrane, (2) ion selectivity, and (3) specificity of pharmacological inhibition. An alternative mechanism that has been proposed is the simultaneous activation of NaCl and KCl symporters; however the KCl symport has been shown to be hypotonically activated (Lauf, 1985), is relatively insensitive to loop diuretics and has an anion preference of $Br^->Cl^-$, the reverse of the Na-K-2Cl cotransport system (MacLeod, 1990).

Studies in small intestinal crypts

We have previously demonstrated that small intestinal crypts undergo complete RVD in response to a sharp decrease in the osmolarity of the bathing medium (O'Brien et al., 1991); and that intact crypts recover their volume after secretagogue-induced volume decrease (SVD) upon washout of agonist (SVI). Measurements of crypt volume after exposure to hypertonic (133% original osmolarity) or hypotonic (67% original osmolarity) solutions are close to those expected for an ideal osmometer, suggesting that estimating crypt morphology to be cylindrical gives a good approximation of crypt volume. Hyperosmotic shrinkage was induced by mannitol addition to ensure that changes in cell volume did not occur as a result of alterations of the ionic composition of the bathing medium. However, it has not been demonstrated whether small intestinal crypts have the capacity to recover their volume in response to hyperosmotic shrinkage

(RVI) or in the maintained presence of secretagogue. In addition, the mechanism of SVI, and RVI, if it occurs, have yet to be elucidated in the crypt.

The % recovery rates per minute in the crypt in response to isosmotic (100nM VIP) and anisoosmotic (133% osmolarity) shrinkage are 4.6±0.8%/min and 4.2±0.8%/min respectively. This suggests that although the rate of SVI appears to be more rapid than RVI, notably in the first 4 minutes after agonist washout, we do have the resolution to answer this question here. The removal of all but 4.2mM extracellular Na^+ by replacement of NaCl with N-methyl D-glucamine chloride is associated with a rapid 5% decrease in cell volume. This is probably due to the depletion of cellular Na^+ that is known to occur through Na^+-K^+ ATPase activity in Na^+-free medium when Na^+ uptake by Na^+-H^+ exchange and/or Na-K-2Cl cotransport is prevented. The gradual reduction in cell volume that occurs in Cl^- free medium might be explained by a reduction in tonic Cl^- dependent Na-K-2Cl cotransport activity. Alternatively, the volume reduction might result from an increased rate of Cl^- efflux from the cell through conductive pathways due to the increased electrochemical gradient resulting from extracellular Cl^- removal. However, failure to observe RVI in the absence of extracellular Na^+ or Cl^- could be interpreted as due to a lack of glucose in the medium, which would provide a metabolic substrate for transport activity (e.g. acetate dependence of Iso-Volumetric-Regulation in renal proximal tubules (Lohr, 1989).

Bidirectionality of cotransport and RVD-RVI

It is interesting that RVI occurs in small intestinal crypts in response to hyperosmotic shrinkage, but not in other cell types (e.g. pancreatic ß-cells). In pancreatic ß-cells both RVD and RVI following the replacement of hypotonic medium with isotonic medium, are effectively inhibited by 1mM furosemide (K.G. Engstrom et al., 1991). RVI did not occur in response to challenge with a medium made hypertonic with the addition of either 50mM NaCl or 200mM sucrose but a slow RVI did occur after prolonged exposure to hypotonicity when the cells were returned to original isotonic bathing medium. This may be due to differences in the gradient for inwardly-directed ion cotransport resulting from an increase in the final intracellular activities of ions after cell shrinkage, which in turn depends upon differences in intracellular ion activities between cell types under isotonic conditions and the magnitude and nature (i.e. elevation of extracellular NaCl or addition of membrane-impermeant osmolyte such as mannitol) of the hyperosmotic "load" placed upon them.

The dependence of crypt RVI and SVI upon extracellular Cl^- ions and upon an inwardly directed Na^+ gradient is consistent with Na-K-2Cl cotransport and NaCl cotransport, as well as with Na^+-H^+ exchange if a Cl^--HCO_3^- exchanger is functioning in parallel. Only a dependence on extracellular K^+ concentration can distinguish between these 3 transport processes. It has been reported that RVI in Ehrlich Ascite tumour cells is highly dependent upon extracellular K^+ ions; RVI being abolished at less than 2mM extracellular K^+, at which concentration cells lose KCl and shrink. The inhibition of both RVI and SVI in the crypt by low concentrations (1m M) of the loop diuretic inhibitor bumetanide and the insensitivity of the volume recovery to either the addition of amiloride (1mM) or ethylisopropylamiloride (40m M), at concentrations known to inhibit both the sensitive and insensitive Na^+-H^+ exchanger subtypes completely is indicative of a Na-K-2Cl mechanism (Na^+-H^+ subtypes: a predictive review, Clark and Limbird, 1991). However, our results are in contradiction to the data of Hamilton and McLeod who reported that RVI in a dissociated preparation of guinea-pig jejunal crypt enterocytes was blocked by amiloride but not 10m M bumetanide. A bumetanide-sensitive Na-K-2Cl cotransport activity has been reported to be present in intact crypt units isolated from rat duodenum, measured by $^{86}Rb^+$ uptake (McNicholas, 1992), but the mechanism of its regulation has yet to be elucidated.

If we simplify by assuming that intracellular ion activities vary in proportion to the change in volume induced in a perfect osmometer then we can predict the initial variation in intracellular activities of individual ions when the cell is exposed to solutions of varying tonicity. If the ß-cell is exposed a solution of 163% original tonicity (by the addition of nonelectrolyte), then we might expect the intracellular ion activities to increase proportionately to the shrinkage in volume. The increased intracellular Na^+, K^+ and Cl^- concentrations that result might effectively reverse the direction of net ion co-transport which is determined largely by the transmembrane gradients of these transported ions. A similar exposure to medium of 163% original osmolarity may similarly inhibit RVI in the small intestinal crypt, in which case the gradient for net cotransport still appears to be inward following exposure to a medium 133% of original osmolarity. This argument may explain why RVI occurs in ß-cells pre-equilibrated in hypotonic medium upon exposure to medium of original osmolarity but not upon direct exposure to hypertonic media. If we assume initial intracellular ion activities in the ß-cell to be 100; 60 and 10 mM for K^+; Cl^- and Na^+ respectively with a relative volume (RV) of 1.0 then exposure to medium of 167% relative osmolarity (RO) will result in an RV of 0.6 and initial intracellular ionic activities of 167; 100 and 17 respectively. However if the cell is exposed first to medium of 83% RO, resulting in an RV of 1.2 and initial intracellular ion activities of 83; 50 and 8 and allowed to undergo RVD until the RV returns to near 1.0; then assuming that K^+ and Cl^- ions are lost in equimolar proportions to restore initial volume we may assume that only small changes in intracellular ion activities subsequently occur. If "isotonic" (relatively hypertonic) medium is restored, a rapid initial return to the original RV of 0.83 will be accompanied by the restoration of intracellular ion activities of 100; 60 and 10mM respectively; considerably less than those predicted to result from exposure to 167% hypertonic medium, thus maintaining a gradient favourable to inwardly-directed cotransport. We might conclude that cells that undergo RVI in response to the hypotonic-isotonic protocol but not on exposure to hyperosmotic media may still effect recruitment of additional cotransporters but net ion uptake may be prevented by a reversal of the transmembrane ionic gradient for uptake. In duck erythrocytes if hyperosmotic conditions are established by the addition of impermeant molecules, rather than Na/KCl, without supplementation of extracellular K^+ above 5mM, then RVI does not occur, as the sum of the chemical potential gradients for Na^+, K^+ and Cl^- are close to zero at low $[K_o]$. However this does not appear to be the case for intestinal crypts wherein the addition of mannitol is adequate for complete RVI to be induced, possibly reflecting a comparatively low resting intracellular concentration of Cl^- and/or K^+ ions in crypt enterocytes.

Comparison of volume regulatory mechanisms in small intestinal crypts

We have demonstrated that crypt RVD and SVD are driven by the parallel efflux of K^+ and Cl^- ions from the cytosol in response to cell swelling and increases in the adenylate cyclase or phospholipase C mediated signalling pathways respectively. Whether the increase in ionic conductance in RVD is due to membrane distention,

Ca^{2+} influx or another signalling pathway (e.g. lipoxygenase metabolic products) has yet to be elucidated. As we have demonstrated that the crypt resting membrane potential is dominated by a basolateral K^+ conductance it may be reasonable to speculate that it is the activation of anionic permeability pathways that is rate limiting for KCl efflux in RVD and SVD, however this remains to be demonstrated directly. If we assume in the resting crypt that steady state volume results from an equilibrium between ion uptake and ion efflux pathways, then for SVD and RVD to occur then the rate of ion loss through permeability pathways must exceed the rate of ion uptake. However, we have demonstrated previously that the crypt membrane conductance is elevated for the duration of secretagogue application. Therefore if, after 8 minutes, a new steady state is maintained at a reduced cellular volume after isoosmotic shrinkage, we can, therefore, argue that the rate of ion uptake must also be increased for this second steady state to occur. This must require either an increase in the number of ions taken up per transporter or an increase in the number of transporters active in the membrane. If K^+ or Cl^- permeabilities are reduced upon secretagogue washout, then the volume loss is recovered as the rate of ion uptake now exceeds the rate of ion loss. In eccrine clear cells methacholine-induced shrinkage is reversed upon agonist washout, but in the continued presence of high concentration of agonist (3m M) most cells (77%) exhibited a gradual volume recovery, whilst 23% showed a maintained shrinkage for the duration of agonist addition (Suzuki et al., 1991). The SVD, which was as much as 30% of resting volume in some cells is of similar magnitude to the mean shrinkages induced by VIP (25%) and carbachol (33%) in the crypts.

It is proposed that the continued progressive shrinkage observed in pancreatic ß-cells in response to the addition of 200mM sucrose is due to a net ion efflux via a furosemide-sensitive Cl^--cation cotransport mechanism (Engstrom et al., 1991). In Chinese hamster ovary cells (CHO cells) hyperosmotic shrinkage activated bumetanide sensitive cotransport, but this did not contribute to RVI, as this resulted in the stimulation of Na-K-2Cl efflux (Rotin et al., 1989).

Bidirectionality is a well-characterised property of Na-K-2Cl cotransport (O'Grady, 1987), consequently at low extracellular K^+ (<0.2mM) the system operates in efflux and cells shrink. This might explain the progressive shrinkage in hypertonicity seen in the crypt when bathed in low extracellular Na^+ or Cl^-. In fact observations from a wide variety of non-intestinal cell-types indicate that intracellular Cl^- must be below a threshold level before shrinkage can activate RVI, a situation created by the RVI after RVD protocol; although this is not necessarily the case for intestinal crypt or villus enterocytes which can recover from a direct anisoosmotic load produced by a solution 154% of original osmolarity.

Role of conductances in RVI

The cell membrane potential was monitored in experiments in which the osmolarity of the bathing medium was increased from 300 to 400 milliosmoles by the addition of 100mM mannitol. An initial depolarisation (mean $10\pm2mV$) of 1-3 min duration was associated with a small apparent increase in membrane conductance ($\approx 10\%$). Within 10-20 minutes of increasing bath osmolarity E_m reached a new steady state $7.6 \pm 1mV$ more hyperpolarised than initial E_m. This hyperpolarisation was slowly oscillating and gradual and was accompanied by a large progressive increase in membrane conductance. The hyperpolarisation phase but not the depolarisation phase of the response to hypertonicity was reversibly inhibited by 50m M bumetanide.

The loss of cytosolic water resulting from an increase in bath osmolarity would increase the intracellular activities of Na^+, K^+ and Cl^- ions as well as increasing the concentration of non-permeable organic osmolytes. As intracellular $[Cl^-]$ increases then E_{Cl} would become more positive, depolarising E_m. Alternatively, the depolarisation could be interpreted by an influx of Na^+ ions or by the inhibition of a membrane-tension dependent K^+ conductance, but the latter would not explain the slight increase in membrane conductance consistently seen. RVI in aortic endothelial cells is associated with a large initial increase in cell Na^+ activity within the first 5 minutes which then recovers towards the pre-shrinkage value (O'Neill et al AJP 1992). Depolarisations have been reported in response to hypertonic stress in intestinal 407 cells (Okada NIPS Vol4 Dec 1989 238-242) and gallbladder epithelial cells (L. Reuss PNAS 82: 6014-6018; 1985). Similar depolarisations have been reported in response to the hypo-iso protocol in Int.407 cells and in MDCK cells (Roy et al J.Memb.Biol. 100: 83-96 1987), which in the Int.407 cells has been attributed to a Na^+ conductance. Such a depolarising entry of

Na$^+$ ions would favour Cl$^-$ ion entry via Cl$^-$ channels activated by prior hypotonic shock, and this is proposed to be the mechanism of RVI in MDCK cells under these conditions.

The hyperpolarisation phase, but not the large conductance increase, could be explained by a reduction in Cl$^-$ permeability. However the gradual hyperpolarisation of membrane potential could be reasonably explained by a progressive accumulation of K$^+$ ions intracellularly due to the parallel operation of the bumetanide-sensitive co-transport and Na$^+$-K$^+$ ATPase activities, the latter possibly activated by an increase in intracellular Na$^+$ activity. The large increase in conductance could be explained by an increase in K$^+$ recycling through the constitutively active basolateral K$^+$ conductance as the electrochemical gradient for K$^+$ exit would be expected to increase with its intracellular activity. This would also account for the movement towards E_K. In rat pancreatic ß-cells exposure to medium made hypertonic by the addition of 200mM sucrose is associated with a 40% increase in ^{86}Rb$^+$ efflux, and this increase is abolished by the addition of 1mM furosemide (K.G. Engstrom et al BBA 1991). Similar observations have been made in cultured HeLa cells where the addition of mannitol to the bathing medium resulted in an increase in loop-diuretic insensitive ^{86}Rb$^+$ efflux persistent with the hypertonic stimulus.

Mechanism of transporter modulation

The mechanisms by which cells sense volume loss and transduce these signals to the modulation of ion transport remains to be elucidated. Two hypotheses of how volume sensitive transport mechanisms are regulated predominate; (i) a mechanical strain sensor in the cell membrane or its associated cytoskeleton modulates the transporter and (ii) that cell volume changes alter the intracellular concentration of a soluble moiety initiating events that lead to transporter activation. Okadaic acid activates Na-K-2Cl cotransport in human red cells even in the absence of cell shrinkage (Pewitt et al J.Biol. Chem. 265, 20747-20756 1991). The principle is that okadaic acid by blocking phosphatase activity leads to the progressive phosphorylation of some regulatory component by kinase activity, and furthermore, protein kinase inhibitors inactivate cAMP-induced cotransport. Thus stimulation by cell shrinkage increases cell phosphorylation relative to dephosphorylation so that fractional phosphorylation increases and the cotransporter is progressively activated. In osmotically shrunken lymphocytes increased protein and phospholipid phosphorylation occurred, with an associated increase in turnover of phosphoinositides that was not as a result of PKC activation, and might possibly be due to the activation of a phosphoinositide kinase (Grinstein, 1986).

A greater stimulation of Na-K-2Cl cotransport was observed in aortic endothelial cells when shrunken isoosmotically than when shrunken hyperosmotically (O'Neill et al, 1992). An increase in bumetanide-sensitive K^+ influx was observed upon hypertonic shrinkage in the aortic endothelial cell. Increased bumetanide-sensitive K^+ influx was associated with an increased number of [^{3}H]bumetanide binding sites rather than an increased influx per binding site in shrunken cells.

O'Neill et al (AJP 262 C436-C444, 1992) have proposed that changes in cell volume alter the number of functional cotransporters in the membrane either by the recruitment of preformed cotransporters to the membrane or due to the activation of latent transporters incapable of either transport or bumetanide binding. They also suggest that cotransport activity is regulated by trans-inhibition by intracellular Cl^-, first proposed to occur in the squid axon (Breitweiser et al., 1990). Interestingly Na-K-2Cl cotransport is also activated by raising intracellular cAMP in avian red blood cells in the absence of cell shrinkage (Kregenow, 1981) and by vasoactive peptides in endothelial cells (O'Donnell, 1989). However Na-K-2Cl cotransport was stimulated in Chinese Hamster Ovary cells but did not mediate a net influx of ions. RVI in these cells was mediated instead by Na^+-H^+ exchange (Rotin et al., 1989). MacLeod and Hamilton concluded that in villus enterocytes under isotonic, isoosmotic conditions that the Na-K-2Cl cotransporter is active, but that at equilibrium does not contribute to the maintenance of steady-state volume and thus bumetanide has no effect on cell volume.

Cell shrinkage has been reported to stimulate Na^+-H^+ exchange via the PKC-independent activation of a G-protein in barnacle muscle fibres (Davis et al., 1992). It will be interesting to determine if G-proteins are involved in volume shrinkage signal transduction in other cell types, including those in which RVI is mediated by transport systems other than Na^+-H^+ exchange.

In small intestinal crypts the mechanism of activation of ion uptake in RVI and SVI in crypts in response to hypertonic or secretagogue challenge appear to be the same. Secretagogues induce volume decrease under isoosmotic conditions, in which intracellular ion activities may change appreciably, particularly if a non-selective cation conductance is activated. Hypertonic (anisoosmotic) challenge will result in a large associated increase in intracellular ion activities, although its effects upon phospholipase C and adenylate cyclase-mediated signalling pathways are unknown in the crypt. If SVI induced upon carbachol washout is also mediated by Na-K-2Cl cotransport it would seem less likely that increases in intracellular Ca^{2+} or cAMP are the specific signalling pathway involved in the mechanism of signal transduction.

There appears to be no apparent correlation between the physiological function of a cell type and the transport mechanism (i.e. Na-K-2Cl cotransport or Na^+-H^+ exchange) employed in RVI. Of all cell types capable of RVI that have been characterised above, those that perform a tissue compartment interface/ barrier function (i.e. endothelial, epithelial and renal cells) show no distinct tendency to fall into either category. However fewer cell types mediating RVI principally by a Na-K-2Cl cotransport system (3/8) are able to do so without prior exposure to a hypotonic medium (the RVD before RVI protocol) than those employing a Na^+-H^+ exchange mechanism (6/8), possibly reflecting a greater capacity of the Na^+-H^+ and Cl^--HCO_3^- antiports to accumulate Na^+ and Cl^- ions against a transmembrane gradient, driven by the carbonic anhydrase and the Na^+-K^+ ATPase activities. A study of the rates of RVI in various cell types again shows no clear difference between cell types or transport mechanism employed. There is, however, a considerable variation in RVI rates from the slowest (mouse pancreatic ß-cell 0.33% change in volume per min) to the fastest (intestinal 407 cell line 9.5%/min). In most cell types tested the rate of RVI is commonly between a 1 and 2% change in cell volume per minute.

Future directions

The molecular mechanisms coupling volume loss to the activation of ion transport in epithelial and non-epithelial cell types is essential to our understanding of cell volume homeostasis. Future strategies may involve the cloning of the epithelial Na-K-2Cl cotransporter and its introduction into cell types known not to express it; or the introduction of mRNA into the Xenopus oocyte expression system and determining the effect of various stimuli upon cotransporter mediated ion fluxes.

If the Na-K-2Cl cotransporter is indeed activated in response to cell shrinkage, it might be reasonable to assume that this could be detected as an increase in the bumetanide sensitive component of $^{86}Rb^+$ uptake into crypt units in response to exposure to secretagogue or hypertonicity. A large increase in the initial rate of bumetanide sensitive $^{86}Rb^+$ influx was induced by hypertonic shrinkage in guinea pig jejunal villus cells (MacLeod and Hamilton, 1990).

If the secretagogue induced shrinkage of isolated crypt diameter prevents a visible expansion of lumenal volume, then in the intact epithelium where the crypt is firmly anchored to the basement membrane SVD might well be associated with a discernible expansion of the lumenal space, which might represent an adaptive mechanism for the increased fluid flow evoked in intestinal secretion.

APPENDIX I

Cystic Fibrosis: the search for an alternative chloride channel

European sailors of the Renaissance searched relentlessly for the elusive North-West passage, a mystical channel which would allow them to avoid the savage seas of the Southern Capes and provide a more direct trade route to the Orient. Their search for the North-West passage was driven both by necessity and by the prospect of economic gain. This perhaps provides an appropriate analogy for the search for a treatment for Cystic Fibrosis (CF), the most common lethal inherited disorder amongst Caucasians. Cystic Fibrosis is a disease which is characterised by a deficit in the pathway for the regulated movement of Chloride (Cl-) ions across a number of epithelia, impairing the hydration of the airways and intestines. Two faulty copies of the CFTR gene are necessary to produce a CF phenotype which is associated with the malabsorption and maldigestion of food and nutrients from the intestine, and the development of a dry, infected lung. The failure of the intestinal crypt epithelium and pancreatic ducts to secrete fluid and electrolytes causes intestinal blockage (meconium ileus), which was originally the primary cause of death in children affected by CF (O'Loughlin & Grant Gall, 1989). The voyage of discovery seeks an alternative Cl- conductance pathway which may bypass these troubled waters.

The frequency of the CF allele may be as high as 5% in the Caucasian population, leading to a CF incidence in the region of 1 in 2000 of all Caucasian births, an extraordinarily high prevalence for such a lethal mutation. The explanation as to why CF is the world's commonest lethal heritable condition rests in the advantage of carrying just one copy of the CF gene. Just as the sickle cell anaemia heterozygote (one faulty gene) is believed to afford protection against the malarial parasite, so a single copy of the CF mutation may protect against the entry of the typhoid bacterium across the intestinal epithelial lining, and was therefore selectively favoured during the great European typhoid epidemics (Pier et al., 1998), and may also protect against bronchial asthma (Schroeder et al., 1995). The most frequent CF mutation, with a frequency of around

70% of all CF mutations, is the so-called 'delta' F508 mutation which results in the deletion of a single phenylalanine amino acid residue from position 508 of the CF protein. In contrast to other CF mutations, which result in the insertion of a defective Cl- ion channel into the luminal (apical) membrane, this ΔF508 mutation results in the failure of the efficient transfer of the CF gene product to the apical (luminal) membrane of the epithelial cells which line the airways and the crypt regions of the small and large intestines. The consequence for the phenotype is the same, whether the CFTR Cl- channel is dysfunctional or entirely absent from the membrane, i.e. there is a deficit in the apical Cl- conductance in these epithelia which normally increases in response to phosphorylation by the action of hormones and neurotransmitters which trigger fluid secretion. Many CF mutations thus block Na^+ and Cl- secretion in response to such secretory modulators (and also to bacterial toxins which cause fluid hypersecretion, or diarrhoea).

Over many years the mechanisms of epithelial secretion have been studied in great detail to find potential pharmaceutical and gene therapeutic strategies towards the treatment of CF (table 1).

Table 1. Primary epithelia affected in Cystic Fibrosis

Tissue affected	Pancreas	Intestine	Sweat gland	Airways	Reproductive organs
Pathology	Maldigestion	Malabsorption/ meconium ileus	salty sweat	Dry airways/ infection	Sterility
Epithelia affected	Exocrine duct cells	Crypt epithelium	Sweat duct	Bronchial epithelium	Vas Deferens/various
Potential Therapies	Enzyme replacement	None	None	UTP + amiloride aerosol/ gene therapy	None

The central role of Cl- channels in CF was established in 1983 when Paul Quinton elegantly demonstrated that reabsorptive sweat duct cells have an abnormally low Cl- ion permeability, that is to say that the membrane of CF duct cells does not allow Cl- ions to cross readily. This provided an explanation as to why CF patients had increased concentrations of NaCl in their sweat owing to decreased salt reabsorption. A deficit in membrane Cl- permeability in response to hormones which increase the levels of an intracellular messenger called cyclic AMP has since been widely demonstrated in other fluid and salt secreting epithelia affected in CF. To understand this defect, we must first understand how epithelia transport electrolytes and water.

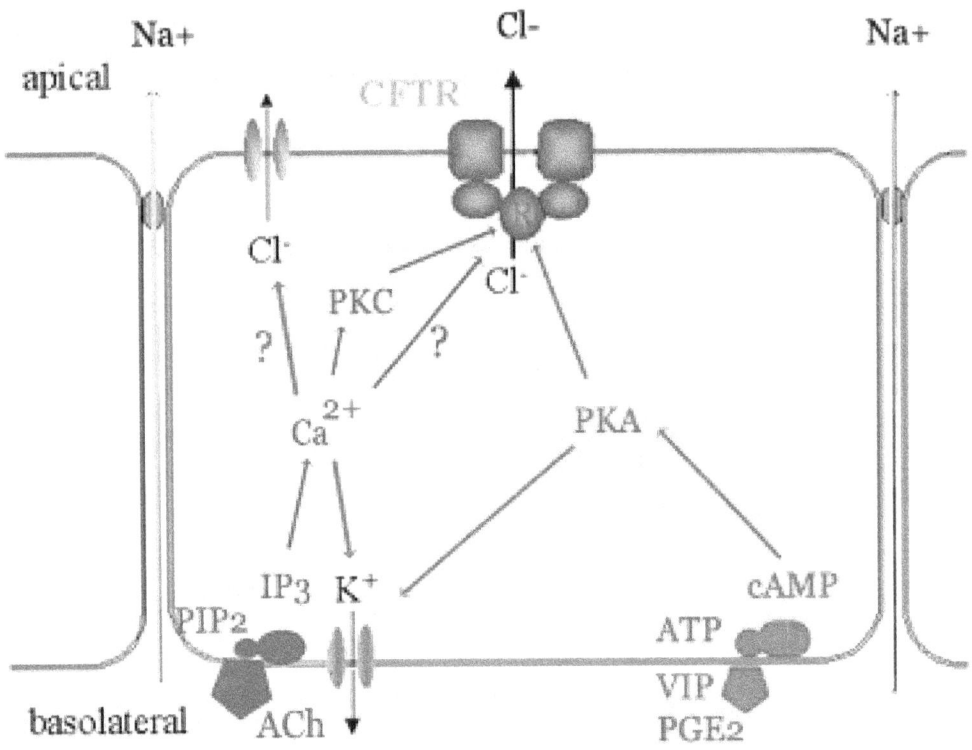

Epithelial cells form a single layer joined together by tight junctions that separate the membrane into two domains, an apical one facing the duct or lumen, and a basal (basolateral) one facing the cellular tissue which is bathed by small blood vessels. Na+, Cl- and K+ ions are taken up by a co-transport mechanism in the basal membrane, and the K+ and Na+ ions are actively recycled across the basal membrane by energy-dependent transport mechanisms, so that concentrations of K+ and Cl- are maintained within the cell which are higher than would be predicted if they distributed themselves freely and passively across the membrane under the influence of the concentration gradient and the transmembrane potential. When hormones interact with membrane located receptors they stimulate an increase in the levels of second messengers such as cAMP or Ca^{2+} in the cell. These second messengers activate channels that increase the rate at which K^+ ions leave the cell across the basolateral membrane and Cl- ions leave the cell via channels that are present in the apical membrane. This results in a transepithelial potential gradient being established across the entire epithelial cell layer because of the negative charge carried into the lumen by Cl- ions. Positively charged Na^+ ions are obliged to follow the negative charge gradient across the epithelial cell layer through cation selective tight junctions, and water follows the NaCl, driven by the osmotic gradient. In CF, the Cl- channels normally activated by cAMP and PKC are not present or function abnormally, leading to an inability of the epithelium to secrete salt and water adequately (Fig.1).

Strategies for overcoming the CF deficit

In 1988 Mike Gray and co-workers classically showed that cAMP-activated Protein Kinase (PKA) activates small Cl- channels in the apical membrane of the pancreatic duct cell. This occurred around 1 B.C., or one year Before Cloning of the CFTR gene, or Cystic Fibrosis Transmembrane conductance Regulator, was announced in 1989 by Francis Collins and his team. Since researchers discovered that CF was due to a Cl- channel defect they have tried two strategies to correct the symptoms of the disease. One of these has been to use gene therapy to introduce a good copy of the CFTR gene into the CF lung either with a vaccinia virus or with membrane microspheres, called liposomes. The defect in genetically engineered CF mice has already been successfully corrected in the short-term with liposomes carrying the correct copy of the DNA (Hyde et al., 1993). An alternative strategy is to activate other Cl- channels in the apical membrane thereby bypassing the CF deficit. Boucher and Stutts (1992) showed that extracellular ATP acting as a hormone activates alternative Cl- channels in CF lung tissue thus bypassing the CF deficit. Soon after, the crypt regions of the small and large intestines were identified as the regions of the intestinal epithelium affected in Cystic Fibrosis (Walters et al., 1992; Trezise & Buchwald, 1992). The search for an alternative apical Cl- channel in the intestinal epithelia was underway.

Having established that many mutations may lead to mild or severe Cystic Fibrosis, and that a deficit in the capacity to secrete chloride ions is predominantly responsible, the burning question in CF remains as to how best to repair the deficit. Whilst researchers in the 80's and early 90's dreamed that drugs might be used to bypass the deficit, the last decade has seen the explosive advent of gene therapy, driven in all fields by the ground-breaking work of the CF teams. Major barriers preclude the successful use of gene therapy in the treatment of CF. The first problem is that not all the affected cells take up the recombinant DNA that contains the 'good' version of the CF gene. In fact transfection efficiencies (the proportion of cells which take up and express the corrective DNA construct) are reported to be as low as 4% even using optimised nebulizing sprays to disperse the vector throughout the airway. Clearly most or all the cells should preferably take up the reforming DNA. This is clearly not merely a problem of distribution in the lung, as transfection efficiencies of cells in culture using a genetically modified strain of vaccinia virus and lipofectamine were as low as 10%. Ten years later using lipofectamine and plasmid vectors, transfection efficiencies had not improved. When cell cultures are treated with fluorescently labelled "naked" antisense DNA, some 10-20% of cultured cells take up the alien DNA in abundance, whilst most show no signs of uptake. This all-or-none uptake and expression pattern is most likely due to the fact that most cells are only receptive to taking up foreign DNA at a certain stage of the cell cycle, possibly during mitosis (cell division). As only a certain percentage of cells in the airway are at this stage of the cell cycle when treated, transfection efficiencies will be low, even before distribution is taken into account.

The second problem is one of turnover. The cells of the airway and intestines are constantly being shed and passed out of the system as we cough or defecate. Hence within a few days, those cells which have been successfully transfected with the corrective DNA, as indeed many cells are, will be lost along with the processes of wear and tear and renewal. Hence any gene therapy protocol may have to be administered almost daily to succeed, and the toxicological consequences of daily high dose gene therapy are a potential cause for concern. The third major problem is that when we introduce foreign DNA into whole organisms we enter a difficult domain. When cells detect single-stranded DNA or RNA or are transfected with a virus, they recognize the DNA or RNA as foreign and enter into programmed cell death (apoptosis). Whilst this is not a bad thing in gene therapy within the realm of cancer, it is not a desirable side effect in the treatment of CF. Further, although people have proposed introducing modified or attenuated viruses in the treatment of AIDS through immunization, the same concerns regarding the use of genetically modified viruses apply in gene therapy. This is to say that even mild strains of a virus may have unpredictable consequences for human health. They may cause new disorders, for instance by over sensitising the immune system, or by inducing latent viruses within the genome to replicate (multiply), or else by recombining in new and unforeseen ways with other viruses. In effect, we may be accelerating the viral evolution by the genetic recombination of new viral strains at a rate that does not take place in the natural world. A single laboratory can create thousands of new viral sequences every year by cutting, pasting, recombining and mutating existing ones. Whilst stringent safeguards do exist for their production and dissemination, the consequences of introducing any new strain of virus into the human body cannot be entirely foreseen.

So where does this leave prospects for CF therapy? Certainly, gene therapy and embryonic stem cell treatment are still in their infancy, and will almost certainly prevail through the irresistible forces of human will, ingenuity and the seemingly limitless resources available for medical research. For the next ten to twenty years however we should perhaps pay attention to the simple and ingenious work of Stutts & Boucher who showed us that UTP, in combination with amiloride, can bypass the CF deficit through the activation of an alternative Cl^- conductance in the airway. Alternate Cl^- conductances most likely exist within the pancreas and intestine, and these may too provide potential therapeutic strategies in the shorter term until stem cell and gene therapy research provide safer and more durable treatments for CF. So for the immediate future, pharmacology will remain our primary weapon in the treatment of CF, and hence we must return our attention to the hunt for alternative Cl^- conductances in epithelia and understanding how these are regulated.

The search for alternative Cl- channels

Fundamental to our understanding of the physiology of small intestinal secretion is the location of the Cl- conductance pathway(s) to either the apical or basolateral membrane, and the determination of whether the second messenger pathways regulating changes in membrane Cl- conductance converge upon the same Cl- channel, or act via distinct Cl- channel pathways. The volume-activated Cl- conductance of Necturus enterocytes has been shown to be located in the apical membrane domain (Giraldez et al., 1988), whilst in isolated colonic crypts volume recovery following cell swelling been shown to be mediated by the activation of basolateral Cl- channels by a mechanism that is dependent upon extracellular Ca^{2+} and prevented by inhibitors of the lipoxygenase pathway (Diener et al., 1992). The location and mechanism of regulation of the constituent ion channels which underlie the various Cl- conductances present in the small intestinal crypt epithelium remains to be established and may be of potential importance in the development of therapeutic approaches for the treatment of CF. In the CF intestine both cAMP- and Ca^{2+}-stimulated Cl- secretion are defective (Taylor et al., 1987; Berschneider et al., 1988), whereas Ca^{2+}-stimulated Cl- secretion appears to be unaffected in CF airways (Wagner et al., 1991; Widdicombe, 1986). Thus although Ca^{2+} ionophore appears to evoke electrogenic secretory processes across normal human jejunum, the effect upon Cl- conductance may be mediated by the action of PKC rather than by Ca^{2+} and its intracellular mediators, since the ionophore-induced increase in short-circuit current does not occur across the human CF jejunum (O'Loughlin et al., 1991). The body of evidence supports the contention that PKG, PKA, and PKC all activate the intestinal CFTR, and unless CF mutations also result in the defective regulation of other types of Cl- channel (Gabriel et al., 1993), there is conflicting evidence for the presence of an alternative Cl- conductance pathway activated by secretagogues in the small intestinal epithelium. In the CF airway epithelium, the Cl- secretory deficit has been successfully circumvented by the luminal application of nucleotide triphosphates such as ATP or UTP, (Stutts et al., 1992; Knowles et al., 1991). Thus a better understanding of the nature and the regulation of alternative Cl- conductive pathways present in the apical membrane of the intestinal crypt epithelium is essential if new drugs are to be developed to overcome the secretory deficit in the CF intestine.

Elucidating the Cl- conductances of the intestinal crypt

One fundamental observation from membrane conductance studies in small intestinal crypts is that there is likely a tonic and sustained secretion of fluid and electrolytes from the intestinal crypt compartment (Walters & Sepulveda, 1991). The resting membrane potential of the intestinal crypt, studied in the absence of neurotransmitters or hormones which stimulate cAMP and Ca^{2+} activated secretion, is dominated by a basolateral K^+ conductance (Walters & Sepulveda, 1991), although there is also a substantial apical component which is diminished by Cl- channel inhibitors. Thus the crypt most likely contains Cl- channels other than the CFTR which mediates a resting level of Cl- conductance and a basal level of fluid and electrolyte secretion. As the apical membrane is inaccessible to study in the intact cylindrical compartment of the crypt, studying these channels requires extreme and innovative techniques, such as the reconstitution of apical membrane within artificial membrane bilayers, or the isolation of a highly purified population of single crypt enterocytes by a sequential process of denudation, sieving, dissociation and marker-driven cellular enrichment. Alternatively, a new model for intestinal secretion may be developed, one which overcomes the difficulty of accessing the apical membrane in the intact intestinal epithelium. Transimmortalized mouse intestinal cells (m-ICc12) have been created which maintain a crypt phenotype and form confluent monolayers of morphologically recognisable enterocytes in vitro (Bens et al., 1996). Such a cell monolayer preparation could provide a preparation which would allow simultaneous recording of short-circuit current and apical chloride channel activity in response to agonist stimulation.

Whilst both the cAMP/PKA and Ca^{2+}/PKC (CaPKC) signalling pathways augment secretion across the small intestine by enhancing a basolateral K^+ conductance, both the cAMP/PKA and CaPKC pathways appear to be impaired by the CF deficit, and as both pathways act cooperatively to enhance K^+ currents so as to maintain the electrical driving force for Cl- secretion, it must, therefore, be the Cl- conductance which is rate-limiting for intestinal secretion - a proverbial electrochemical "bottle-neck". The primary question in intestinal secretion is whether both cAMP/PKA and CaPKC pathways act synergistically through the CFTR Cl- conductance to augment increases in membrane Cl- current alone, as the CFTR is activated more potently by PKC and cAMP/PKA than by either kinase alone (Walters et al., 1992), or whether cAMP/PKA pathway activation renders Cl- channels, normally unresponsive to the CaPKC pathway, sensitive to its intracellular mediators. However, even if apical Cl- channels other than the CFTR channel are activated *in vivo*, they are clearly insufficient to bypass the symptoms of CF without clinical intervention. Thus if other Cl- channels are found, their mechanisms of regulation must be fully elucidated if we hope to circumvent the CF deficit.

Of particular contention is whether a rise in intracellular free Ca^{2+} results in the activation of an alternative Cl- conductance. Crude conductance measurements from whole crypts from small intestinal (Walters & Sepulveda, 1991) and colonic crypts (Bohme et al., 1991; Jens Leipziger personal communication), in addition to studies from the intact CF intestine (Taylor et al., 1987; Berschneider et al., 1988; O'Loughlin et al., 1991), suggest the absence of a Ca^{2+}-activated Cl- conductance, which is common in other fluid and electrolyte secreting cells (Cliff et al., 1990; Randriamampita et al., 1988; Wagner et al., 1991). However, volumetric measurements from small intestinal crypts derived from CF mice using the muscarinic CaPKC mobilizing agonist carbachol and crude Cl- channel inhibitors have suggested otherwise (Valverde et al., 1993). Whilst the T84 cell line, a lung metastasis of a colonic carcinoma epithelial cell line, which retains characteristics of both undifferentiated and differentiated lung and colonic epithelial cells, has been reported to express a Ca^{2+} activated Cl- conductance pathway (Barrett & Keely, 2000), this naturally cannot be taken as conclusive evidence in support of either school of thought.

Since the suggestion of Valverde et al (1993) that a putative intestinal Ca^{2+}-activated Cl- conductance might provide a viable means to bypass for the CF deficit, many attempts have been made to characterise it. Gruber et al (1998) established that hCLCA1, the first human member of the family of Ca^{2+}-activated Cl- channel proteins to be identified, is exclusively expressed in intestinal basal crypt epithelia and goblet cells. Gruber and co-workers believe that hCLCA1 likely produces a functional Ca^{2+}-activated Cl- conductance in the human intestine, making for an interesting candidate. The expression cloning of an ileal brush-border (apical) Cl- conductance also led to the isolation of CLCA1, which produces a Ca^{2+} activated Cl- conductance activity when expressed within a heterogenous expression system (Gaspar et al., 2000). Whilst not conclusive, there is now both physiological and molecular evidence for the presence of a Ca^{2+}-activated Cl- conductance in the apical membrane of the intestinal crypt. If such a Ca^{2+} activated Cl- conductance pathway does indeed exist, the principal action of Ca^{2+} mobilising agonists upon membrane potential seems to be dominated by the change in K^+ conductance (Walters & Sepulveda, 1991; Bohme et al., 1991). This would be in agreement with previous results obtained in T84 colonic carcinoma cells, where the muscarinic (CaPKC) agonist evokes Cl- secretion through pre-activated Cl- channels by increasing basolateral K^+ permeability only (Dharmsathaphorn & Pandol, 1986). In other words, if such a Ca^{2+}-activated Cl- conductance (CACC) were to exist in the crypt *in vivo*, and if it were functionally expressed in the apical membrane, would its activation be physiologically or pharmacologically be sufficient to ameliorate the CF deficit? Clearly such an apical Ca^{2+} activated Cl- conductance presents an irresistible target, although the more recent report (Valverde et al., 1993) conflicts with earlier findings suggesting that the CF defect blocks Ca^{2+} induced Cl- secretion (Taylor et al., 1987; Berschneider et al., 1988; O'Loughlin et al., 1991). Whilst species-specific differences are a popular argument to explain away such anomalies, an appealing explanation might be that the CFTR somehow alters the functioning of an alternative apical Ca^{2+} induced Cl- conductance pathway, as has been shown for other conductance pathways in the airway epithelium (e.g. Gabriel et al., 1993).

Alternative candidate Cl- channels

A number of Cl- conductances are believed to exist within crypts or colonic epithelial cell lines. For any of these to provide strong candidates as therapeutic targets they must (a) be proven to be distinct Cl- channels directly regulated by these pathways, (b) be shown to be functionally expressed within the proliferating (secretory) region of the crypt compartment, (c) be localised within the apical membrane of the crypt enterocyte, and (d) they must mediate an increase in Cl- conductance and short-circuit current across the crypt epithelium when activated. As the intestinal epithelium is accessible from both the serosal and luminal faces, any such Cl- channel that has a distinctive pathway of regulation (e.g. a volume activated current) would present an appealing therapeutic target.

Since the small intestinal crypt compartment was first shown to mediate the secretion of fluid and electrolytes (Walters & Sepulveda, 1991; Walters et al., 1992), many other studies have provided both direct and indirect evidence for other Cl- channels in the crypt epithelium which are possible candidates for circumventing the CF Cl- deficit. A ClC-5 chloride channel (gpClC-5) has been cloned and functionally expressed from guinea-pig distal small intestinal epithelial cells, and is homogeneously distributed within the crypt and villus regions of duodenal, jejunal and ileal epithelium (Cornejo et al., 2001). Whilst a specific role in agonist-evoked Cl- secretion seems unlikely, such a Cl- conductance may, for example, play a role in regulatory volume decrease or as a tonic Cl- conductance pathway. Another popular idea is that a Cl- conductance within the crypt epithelium is activated during regulatory volume decrease (O'Brien et al., 1991), then this may be "hijacked" pharmacologically to circumvent the CF deficit. However, measurements of volume and single channel activity in isolated rat colonic crypts following exposure to hypotonic media have shown that RVD is mediated by the parallel activation of basolateral K^+ and Cl- channels (Diener et al., 1992), and if such channels are basolaterally localised in the small intestinal crypt epithelium, then no promising therapeutic strategy appears on the cards. Last, but not least, no discussion of this topic could comfortably omit the release of defensin peptides by Paneth cells (termed cryptdins in mice) into the crypt lumen. Paneth cells are located at the base of the crypt and thus any mediators released from them might be expected to alter crypt physiological function. Defensins form Cl- channel-like activity when applied to apical membranes of airway epithelial monolayers expressing the ΔF508 CF mutation (Merlin et al., 2001) and stimulate Cl- secretion from polarized monolayers of human intestinal T84 cells (Lencer et al., 1997). Thus, the intriguing possibility that defensin release may be induced from the Paneth cells *in vivo* to overcome the CF Cl- deficit has been proposed. Indeed cryptdins 2 and 3 may function as novel intestinal secretagogues, providing a mechanism of paracrine signalling by the reversible formation of ion conductive channels in secretory crypt enterocytes, which are not dependent upon the activation of cAMP or cGMP pathways (Lencer et al., 1997).

A liposomal strategy

One fundamental aspect of the complex pathophysiology of the CF defect has not so far been given due consideration. CF airway epithelial cells provide binding sites for pseudomonas bacteria leading to the destruction of lung tissue, due to abnormally low levels of sialylation and elevated sulphation and fucosylation of apical membrane

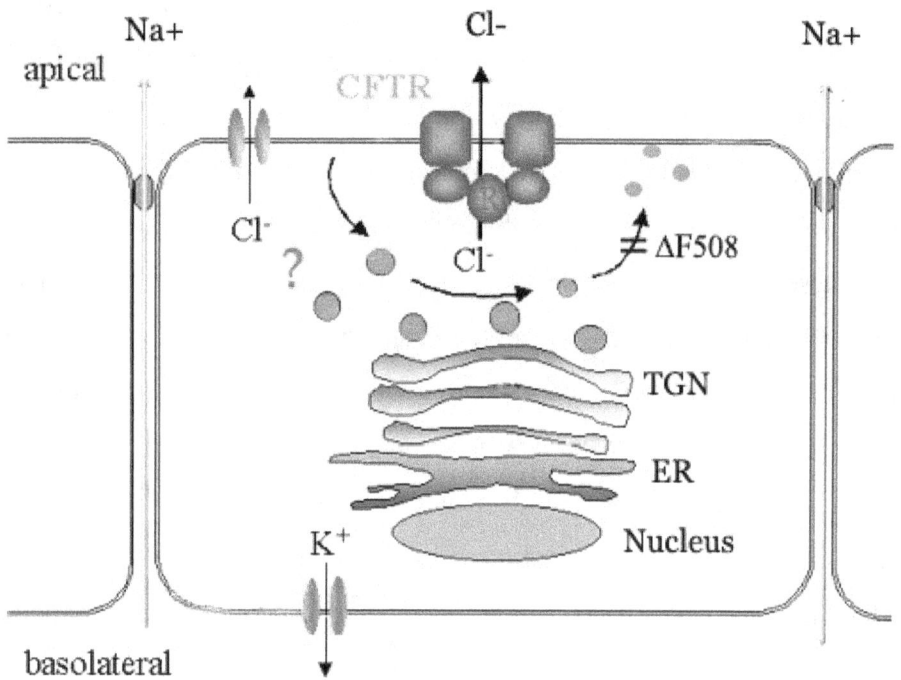

glycoproteins. Barasch and colleagues (1991) suggested that the defective acidification of the trans-Golgi/trans-Golgi network in CF resulted from the diminished CFTR Cl-conductance, leading to defective regulation of glycosylation patterns in secretory granules. Yet more pathology lies below the membrane surface. Wild-type CFTR is known to attain a protease-resistant configuration in an energy-dependent process within the ER, which does not occur in the processing of the mutant ΔF508 CFTR, providing one potential explanation as to why this otherwise functional Cl- channel does not reach the apical membrane in sufficient quantities (Lukacs et al., 1994). Ameen and co-workers (2000) identified CFTR protein in both subapical vesicles, and on the apical plasma membrane of the crypt enterocytes. They further demonstrated that cAMP stimulation produced a fluid secretory response which was associated with a redistribution of CFTR from this vesicular pool to the apical membrane, suggesting that cAMP agonists not only increase CFTR Cl- activity, but also recruit additional CFTR protein to the cell membrane associated with increased fluid secretion (Ameen et al., 2003).

As secretory vesicles containing CFTR (and potentially other Cl- conductances) are recruited to the apical membrane by agonist pathways, and given that the CF deficit interferes with a range of processes including vesicular pH regulation, post-translational modification, and the recruitment and apical membrane insertion of CFTR containing vesicles, it follows that other functional Cl- conductances present within these secretory vesicles might also fail to be functionally delivered to the apical membrane. As the CF deficit associated with the prevalent ΔF508 mutation is not due to the loss of the CFTR Cl- channel function, but is rather a problem with the processing and recruitment and insertion of these channels into the plasma membrane, it might also follow that other Cl- channels which might potentially be employed to bypass the CF deficit similarly fail to be incorporated into the plasma membrane together with the CFTR. Thus one potential therapeutic strategy would be to create therapeutic vesicles, such as liposomes, which may be taken orally to rescue the recruitment of these vesicles into the plasma membrane. Such liposomes may contain a combination of therapeutic components. For instance, they may incorporate docking receptors or integral membrane-spanning antibodies directed against crypt-specific antigens present upon the microvilli of the apical membrane. Such liposomes might feasibly contain a combination of incorporated proteins in addition to gene therapy vectors. For instance, they may contain surface membrane proteins which promote the fusion (docking) of pre-existing CFTR containing vesicles with the crypt apical membrane, or even proteins which otherwise compensate for aberrant pH regulation upon fusion with the CFTR containing vesicles. However such liposomal therapies do not circumvent the need for a better understanding of the Cl- conductances present within the apical membrane of the intestinal crypt.

Conclusions

The presence of apical Cl- pathways distinct from the CFTR within the crypt seems more than a probability. These may ultimately provide a viable means of circumventing the CF Cl- deficit *in vivo*. However, whether these are functionally independent of regulation by the CFTR (which also regulates cationic, ORDIC and potassium channels in epithelia), or are unimpaired in their function by the defective vesicular trafficking and glycosylation patterns characterised by CF is another matter. Liposomes have already been used successfully as gene therapy vehicles to correct the ion transport defect in CF transgenic mice (Hyde et al., 1993), and why should their payload and functional range not be extended towards the treatment of the intestinal symptoms of CF?

APPENDIX J

Schizophrenia: A cyclical and heterogeneous dysfunction of cognitive and sensory processing?

A First Century of Schizophrenia

Schizophrenia remains an enigma that has fascinated the foremost minds of psychiatry and neuroscience for more than a hundred years. At stake is more than just the crucial welfare of the millions afflicted, for schizophrenia research may represent the key to an understanding of the mechanisms by which the brain filters, prioritizes and processes the relentless current of information available from the richness of its internal, social and natural environments.

In 1851 Falvet first described a 'Folie Circulaire' or cyclical madness, and some twenty years later Hecker referred to a 'Hebephrenia', or a silly, undisciplined mind after Hebe, goddess of youth and frivolity (1871). Soon after, in 1874, Kahlbaum referred to both catatonic and paranoid disorders of the mind, the term catatonia describing a movement disorder characterized by a mannequin-like muscle stiffness associated with unusual postures and a pervading fear. Then in 1878 Emil Kraepelin, perhaps auspiciously, combined these various 'disorders' into a single disease entity which he termed ***dementia praecox***, or 'dementia of early onset' reflecting a decline of cognitive processes which he divided into four subtypes - *simple*, marked by slow social decline concomitant with apathy and social withdrawal; *paranoid*, with its attendant fear and 'persecutory' delusions; *hebephrenic* and catatonic, characterized by a poverty of movement and expression.

The inevitable inexactitudes of this emerging science continued with the dawn of the 20th Century when in 1908 Eugen Bleuler criticized the use of the term *dementia praecox,* arguing for an absence of evidence supporting a global dementing process. It was Bleuler who first coined the divisive term '*schizophrenia*' in 1911. Bleuler defined schizophrenia with his four "A's", referring to the blunted _A_ffect (diminished emotional response to stimuli); loosening of _A_ssociations (by which he meant a disordered pattern of thought, inferring a cognitive deficit), _A_mbivalence (an apparent inability to make decisions, again suggesting a deficit of the integration and processing of incident and retrieved information) and _A_utism (a loss of awareness of external events, and a preoccupation with the self and one's own thoughts).

Freud's Paradox

Sigmund Freud, after many years of formative research on the anatomy of the vertebrate nervous system, which culminated posthumously in the publication of a "Project for a Scientific Psychology" (written in 1895), he began:

"The intention is to furnish a psychology that shall be a natural science: that is, to represent psychical processes as quantitatively determinate states of specifiable material particles (*which he termed neurons*), thus making those processes perspicuous and free from contradiction."

However, later that same year Freud wrote in "Studies on hysteria", published in 1895, elegantly summarizing his apparent conflict of ideologies;

"...it still strikes myself as strange that the case histories that I write should read like short stories and that....they lack the serious stamp of science....The fact is that local diagnosis and electrical reactions lead nowhere in the study of hysteria....whereas a detailed description of mental processes....enables me...to obtain at least some insight into the course of that affection". This philosophical dilemma may have been the inspiration for his development of his theory of psychoanalysis. Freud within the same year echoes our dilemma of dreaming for an undisputed biological basis for the operations of the mind, cognitive processes and behaviour, contrasted against the realities of dissecting the "human condition".

The operational definition of schizophrenia

Kurt Schneider listed his 'first rank' features of schizophrenia in 1959 which served as the inspiration for the two guides used in the operational diagnosis of schizophrenia, the ICD-10 and the Diagnostic and Statistical Manual of mental disorders (DSM). The DSM (IV) states that two or more of the following (symptoms), each present for a significant portion of time during a one month period (only one symptom being required if delusions are bizarre or 'auditory' hallucinations are present). A diagnosis of schizophrenia may be made if continuing signs of a disturbance have been present for at least six months, concordant with a social and occupational dysfunction for a significant period of the time since onset, provided other medical conditions and the actions of substance abuse have first been ruled out. This assumption is however challenged below. The classification is summarized below;

Table 1: Major features of schizophrenic phases

POSITIVE SYMPTOMS	NEGATIVE SYMPTOMS
Psychotic episode (displacement from 'reality', inability to separate real from unreal experiences) including; *delusions* (false beliefs/judgment); *hallucinations* (strong subjective perceptions of an object or event which is non-existent that may affect any or all sensory perceptions); *disorganized speech* or *behaviour*; *thought disorder* (cognitive dysfunction)	Social and occupational dysfunction

Lack of motivation, withdrawal, loss of concentration

Blunted or flat affect (loss of emotional tone or reaction)

Inability to articulate |

Table 2: The five principle subtypes in the spectrum of schizophrenic disorders

Subtype	Characteristics
Paranoid	A preoccupation with one or more delusions or frequent auditory hallucinations
Disorganized	Disorganized speech and behaviour and a flat or inappropriate affect are all prominent
Catatonic	Two of the following must be present: A lack of a motor response to a stimulus, excessive motor activity, an absence of speech, peculiar movements and repetitions of words and phrases (echolalia) or another's movements (echopraxia)
Undifferentiated	Symptoms of schizophrenia are present but conditions for other three types are not met
Residual	Absence of prominent delusions, hallucinations, disorganized speech, and grossly disorganized and catatonic behaviour despite continuing evidence of a disturbance

It may be instructive to derive the core features inferred from this operational definition of schizophrenia. Extreme distortions of sensory processing are apparent, with attendant difficulties in screening out various unwanted sensory stimuli or ideations (leading to delusions or hallucinations) which is suggestive of a decreased capacity to filter and process information. The resulting disorganized thought and display of behaviours that do not meet with social expectation are often associated with the development of poor memory and a shortened attention span. The apparent decline in cognitive processing is often reflected in disorganized speech, further suggestive of a deficit in information processing. A decrease in emotional tone and of reaction to social and other external stimuli may parallel a decline in social functioning, emphasizing the importance of the integrity of higher cortical circuits in mediating receptive, productive and appropriate social interaction, or 'successful' human social 'behaviour'. This reductionism and generalization leading to the definition of 'schizophrenia' as one or more related disorders resulting in a disruption of cortical processing and filtering, permits us to correlate these human behaviours with those of animal models from which the putative existence of a biochemical basis for schizophrenia might be tested.

The much maligned and misunderstood schizophrenic

Schizophrenia does not infer, from the literal translation 'split mind', to a dissociation of personality (Jekyll & Hyde) or multiple personality disorder. Rather Bleuler intended it to refer to a split between subjective feeling, or affect, and the thought being experienced. Most schizophrenics have not, contrary to widespread belief, been shown to be unusually prone to violence either normally or following substance abuse. Diagnosed schizophrenics receive discrimination in seeking employment, housing, healthcare and insurance. Thus, schizophrenia is a dysfunction with often severe social consequences.

Are there other conditions that resemble features of schizophrenia?

Psychoactive drugs	Disease
Crack (purified *cocaine*) and ice or **crystal** (pure methamphetamine) cause the positive symptoms of schizophrenia and dysphoria upon withdrawal.	*Schizophrenia-like psychosis of epilepsy* (SLPE), resulting from temporal lobe epilepsy of the left (dominant) superior temporal gyrus, have similar symptoms to schizophrenics but the predominant cause is temporal rather than in the anterior cingulate or frontal lobes, as in schizophrenia
Anti-depressants and serotonin reuptake blockers, such as **ecstasy, Prozac** & **LSD** create hallucinations and false memory.	Prior to this century 10-30% of schizophrenia-like patients had *neurosyphilis*
Chronic alcoholics suffer loss of gray matter, cognitive dysfunction, disorganization of thought and memory loss	*Post-traumatic stress disorder*
Special K, otherwise known as *ketamine*, causes a schizophrenia-like psychosis in healthy individuals and exacerbates the psychotic symptoms in schizophrenic patients, decreasing responsivity to environmental stimuli.	*Sleep deprivation* impairs cognitive performance and causes activity shifts from temporal to parietal cortex on verbal learning tasks (Drummond et al., Nature, 2000).
PCP, or *phencyclidine* otherwise known as '**Angel Dust**', causes both the positive and negative features of schizophrenia	

How may we quantitatively measure schizophrenia?

Psychometric measurements

Thought disorder index in response to questions (TDI)

By a dysfunction in smooth pursuit eye movement (eye tracker)

Evoked potentials

Measuring a deficit in information processing, inhibitory and gating deficits are apparent in both human and animal models of schizophrenia, and this neurological deficit can be revealed by a loss of **prepulse inhibition (PPI)** of the 'startle' reflex. Such PPI deficits in schizophrenic patients are thought to be neural correlates of cognitive deficits such as 'thought disorder' and distractibility. For example, the PPI of the eyeblink component of the startle reflex to a loud noise can be measured by electromyogram, an electrical measurement of muscle activity (Braff et al., Am.J.Psych., 1999).

In addition, both a failure to inhibit the P50 auditory evoked response to repeated stimuli and an increase in latency of the P300 cognitive event-related potential in response to an auditory oddball event (Noldy & Carlen, 1997), which is measured by *electroencephalography* (**EEG**) are features present in diagnosed schizophrenics. It would be instructive to determine what proportion of diagnosed schizophrenics possess these deficits.

Brain Imaging

Changes in the structure and function of the brain may be measured non-invasively by changes in:

- regional blood flow (**functional MRI**)/regional blood oxygen consumption (**BOLD**)
- binding or localization of emitting tracer (**Single Photon Emission Computed Tomography, Positron Emission Tomography**)

o structural changes (**Computed Axial Tomography**)

In what way are the brains of schizophrenics different?

As schizophrenia is believed by many to reflect a disturbance in information processing, and specifically a failure to correlate and integrate contextually appropriate stored material (memory) as a function of sensory input, in other words an inability to effectively relate stored experience to current circumstance. Indeed 'paranoid' and 'non-paranoid' schizophrenic subjects exhibit equivalent performances in tests related to cognitive and intellectual functioning (Zalewski et al., Schizo. Bull., 1998). The general concept of a fundamental cognitive deficit in schizophrenia is unifying and takes into account a broad diversity of symptoms and possible causes by describing a functional consequence rather than defining a specific causality.

What regions of the brain are affected in schizophrenia?

Mesolimbic areas including the amygdala and ventral striatum, believed to be important in imparting emotional 'colouring' to external stimuli, have been shown to be unusually active in schizophrenia, whilst the prefrontal cortex is unusually hypoactive during hallucinations, a symptom of the so-called 'active phase of schizophrenia'. In fact, these patterns of differential activity may be seen by **PET** imaging even without stimulation, suggesting a constitutive state of arousal typically observed in response to threat (Epstein et al., 1999). Brain imaging of responses to non-threatening, negative expressions such as disgust, or threatening facial expressions, such as fear or anger, suggest that a region called the amygdala specifically responds to threatening facial expressions, an area known to be important in the integration of the cognitive and emotional aspects of human behaviour. The amygdala has since been shown to have an enormous influence on dopamine release, thereby emotionally and motivationally 'colouring' a wide range of behaviours. It is suspected by some that the amygdala plays a role in mediating some of the symptoms of schizophrenia.

There is agreement that an area known as the associative frontal (including left dorsolateral pre-frontal) cortex has both reduced blood flow and metabolic activity in 'never-been' medicated schizophrenics, as do the upfoldings (gyri) of the parietal and temporal cortex, all association areas essential in governing the high level cognitive functions implicit in social interaction and language. Further, the prefrontal cortex has been shown to be an area consistently associated with the altered cognitive activity and attentional deficits in schizophrenics, consistent with an area involved in mediating transient working memory and processing information involved in thought, which is impoverished or disorganized in schizophrenics. However, the thalamic and cingulate cortical areas are hyperactive in schizophrenics, regions that are thought to be important in perception and communication.

The hippocampus serves as a thoroughfare for information arriving from sensory areas en route to higher cortical areas for further association and encoding, a seeming crucial junction box both in information processing and learning. In particular the hippocampus has been implicated as central in the formation of memory and learned behaviours, and in particular spatial memory for places and resources. Rats with lesions introduced into their ventral hippocampus showed less time spent in interaction and an enhanced aggression which was not attributable to anxiety. This effect however occurred only in young rats and not in those lesioned after weaning, indicating that damage to cortical integrative circuits sustained before or during the period of learned social behaviours can result in schizophrenic 'asociality' (Becker et al., Psychopharmacology, 1999). Furthermore lesioning of the hippocampus, which processes auditory information, results in a loss of filtering of auditory information presented in the form of paired clicks, as the electrical signal produced by the second click was not substantially attenuated in the lesioned animals. As with the **PPI** test, this suggests that filtering of auditory information is deficient, and may occur at the level of the hippocampus, which mediates, at least in part, the selective filtering of sensory information.

The brain functions, as do its individual units of information processing (neurons), as difference detectors, comparing two inputs, for example sensory input with stored memory, and subtracting the current (sensory) input from the past (stored) experience (which may for example be a recent or evoked memory), and projecting the difference in the form of a (non-linear) output. If there is no difference between a present stimulus and a past stimulus then the pattern may be said to be 'expected' and that there is relatively little that is new to report (from Bilder, Mannarcc, Conference 2000) and thus little information processing capacity is devoted (energy) within higher cortical centers (attention) as the information may not be projected as extensively to higher centers for processing. By this means, only changes in environment 'perceived as significant' are passed forward and unwanted background information is filtered out, although in schizophrenia **it is this filtering that is apparently impaired.**

Andreasen and colleagues (Biol.Psychiatry, 1999) postulate that the diverse symptoms of schizophrenia are due to a single disorder involving the misconnection of neural circuitry within the cortical-thalamic-cerebellar-cortical (CCTCC) circuit, which is critical in the synchronous firing of neuronal centers involved in the smooth co-ordination of mental processes. Andreasen proposes that when the synchrony of the CCTCC circuit is impaired the patient suffers from cognitive dysmetria, and it is this impairment of basic cognitive processes which defines the hallmark of schizophrenia. Hence many different disruptions of this circuit may produce a common phenotype, just as many different ways in which cells may fail to regulate their growth and survival might result in a cancer.

Andreasen has argued that there is connectivity between nodes (or clusters of neurons involved in processing) in the pre-frontal cortex, the thalamic nuclei, and the cerebellum. Any fundamental disruption in the CCTCC circuit may result in a cognitive impairment or dysmetria, associated with a difficulty in prioritizing, processing, coordinating, and responding to sensory information. This, as has been long since demonstrated for the visual system, shows that the brain is not comprised of a series of clearly defined and discrete functional centers ascribed to specific functions such as memory or logical thought, but rather is comprised of diffuse aggregates of functionally associated neuronal clusters, or nuclei, which distribute and dynamically process information in parallel. Thus damage to the delicate microelectronic circuitry of the brain can readily disrupt the highest level brain functions such as language, intelligence, and social behaviour. Difficulties in isolating a cognitive deficit to any one receptor or part of a circuit is difficult, as function is thus devolved, and all neurons are ultimately interconnected. Further, any given neuron may express upon its receiving surface receptors for as many as six or more different neurotransmitters. From these basic observations, there is an implicit divergence of information flow, and the consequences of disrupting signalling via any one type of neurotransmitter may have widespread consequences for information processing across and between many circuits.

A Diaspora of theories of the causation of schizophrenia

A plethora of theories have arisen to explain the deficits associated with schizophrenia including specific or 'global' changes in neurotransmission involving a given neurotransmitter, cell type or receptor, and a host of changes in brain function elicited by agents as diverse as stress, developmental aberrations, and viral infection. These may be summarized as follows,

Changes in neurotransmitter systems

- Dopamine
- Glutamate
- Acetylcholine
- Serotonin
- GABA
- Norepinephrine

Alternate theories

o Miswiring of the brain during development
o An inherited disorder exacerbated by stress and hormonal changes
o A transmissible viral infection
o Perinatal hypoxia
o Autoimmune damage
o A neurodegenerative disorder (such as Alzheimer's, Parkinson's)
o An inherited predisposition interacting with an overload of dietary proteins
o The loss of tissue due to the neurotoxic effects excessive nerve transmission
o Social stressors in urban settings
o Depletion of certain fatty acids in cell membranes
o Dietary Exorphins from milk

The Dopamine hypothesis

The central involvement of a deficit in dopamine function in schizophrenia is suggested by the observation that medications which alleviate the psychosis of schizophrenia such as chlorpromazine (Thorazine) act by antagonizing the actions of dopamine at its receptor, especially the D2 receptor. Bilder proposes that dopamine systems mediate the comparison of observed (perceived) and expected ('normal') patterns of events within resonant cortical circuits, measuring departures from 'normality' as deviations from patterns of expected resonance. By antagonizing dopamine receptors it is possible to block these perceived shifts, and the attendant projections of 'non-existent' perceived events (hallucinations) during the active phase of schizophrenia.

Dopamine receptors are present in the prefrontal cortex (PFC), nucleus accumbens, striatum, hypothalamus and hippocampus, and are believed to mediate the motivational aspects of reward and reinforcement following dopamine (DA) release from projections from the ventral tegmental area into the striatum, nucleus accumbens and PFC which are under hippocampal influence. Further, drugs which increase dopamine by blocking its reuptake by the DA transporter into nerve terminals such as cocaine and amphetamine cause the positive symptoms of schizophrenia, such as increased locomotion, hallucination and other aspects of psychosis at high concentration. Further, in 1995 Csernansky & Bardgett suggested that damage to hippocampal neurons, whose projections impinge upon dopaminergic terminals coming from the midbrain, might explain the anatomical and functional abnormalities in gating of sensory information observed in schizophrenia.

The contention that an excessive state of dopaminergic activity is present in schizophrenia has been demonstrated by Laruelle and colleagues who showed by SPECT imaging that acute challenge with the dopamine uptake inhibitor amphetamine caused a greater, although variable, increase in dopamine release from presynaptic terminals in the striatum of schizophrenics as measured by an increase in the occupancy of D2 receptors in schizophrenic patients relative to 'healthy' patients, indicating some manner of a dysfunction of dopamine regulation in these individuals (Laruelle et al., 1999). However, this could only be observed during the active phase, when psychotic episodes are known to occur. However, there was no difference noted in D2 receptor availability in the absence of amphetamine challenge (or during periods of remission), although amphetamine challenge stimulated a worsening or emergence of positive symptoms, suggesting that schizophrenics exhibit a dysregulation of striatal dopamine release (Abi-Dargham et al., 1998). However, there is a small, but significant increase both in D2 receptor density post-synaptically and in presynaptic DOPA decarboxylase activity (Laruelle, Quart.J.Nuclear Med. 1998).

As dopamine is held to modulate, or 'colour', the tone of excitatory transmission through projections of the frontal and temporal cortex to the basal ganglia (e.g. striatum) and other areas, dopaminergic transmission is thought to be implicated in the deficit in information processing associated with the prefrontal cortex, the PFC having one of the highest concentrations of dopaminergic nerve terminals. It is believed that a high concentration of certain DA receptors interact with glutamate receptors to facilitate memory formation in the PFC (D1?). Professor Goldman-Rakic has argued that the 'derailed' train of thought associated with schizophrenia is due to a deficit in working memory, the executive function of the prefrontal cortex. As the prefrontal areas receive a high concentration of dopamine, and drugs that are effective in treating schizophrenia act on these dopamine receptors, such as clozapine, Professor Goldman-Rakic argues that enhanced dopamine levels 'induced' by clozapine act to improve thinking and memory. Perhaps this is the mechanism by which cocaine and Ritalin (methylphenidate) enhance cognitive performance, attention and memory, by increasing availability of DA in the PFC through an inhibition of reuptake. If so, there must be a fine balance, as cocaine and other dopamine uptake inhibitors also cause the positive symptoms of schizophrenia at high doses.

The Glutamate Hypothesis

Much of the transmission of excitatory information in the brain occurs via the binding of glutamate to its receptors, and, directly or indirectly, the activity of most neurons in the brain are influenced by this excitatory amino acid. The blockade of one specific glutamate receptor, the NMDA receptor, which plays a critical role in the plasticity of nervous connections associated with learning and memory, appears to mimic certain symptoms of schizophrenia. Two of the more popular psychoactive drugs of the 1970's, phencyclidine ('Angel Dust') and ketamine ('Special K') specifically block NMDA receptors and cause hallucinations in humans, as well as stereotyped, repetitive behaviour and social withdrawal in both rats and humans, thereby reproducing both the positive and negative symptoms present in schizophrenia. Ketamine, a dissociative anesthetic, causes a schizophrenia-like psychosis in healthy individuals and exacerbates the psychotic symptoms in schizophrenic patients, decreasing their apparent responsivity to environmental stimuli (Shiigi & Casey, Psychopharmacology, 1999).

In light of these observations, one theory which has been put forward is that schizophrenia results from a hypoactivity of glutaminergic transmission in the brain. A decrease in glutaminergic output from the hippocampus, coupled with the high levels of glutamate receptors present (particularly NMDA) in the anterior cingulate cortex, one of the principal targets of hippocampal glutaminergic output, may underlie some of the diminished cognitive aspects and processing deficits associated with schizophrenia. In contrast, Professor Dan Javitt and colleagues have suggested that treatment with glycine, a co-transmitter essential for proper NMDA functioning, improves the negative symptoms of patients with schizophrenia (Heresco-Levy et al., 1999). Memory and other cognitive deficits in schizophrenic patients may be explained in part by reductions in the transcript (mRNA) for the glutamate receptor subunits NMDAR1, GluR1, GluR7 and KA1 mRNA levels in frontal cortex both in drug-free and drug-withdrawn schizophrenics (Sokolov, J.Neurochem., 1998).

The NMDA receptor has a key function in information processing as a coincidence detector implicated in the molecular basis of learning and memory, as converging and coincident signals must be received by a nerve cell bearing this receptor upon its receiving terminals, or dendrites, for its activation (see Eric Kandel, Principles of Neural Science). Many have proposed that abnormalities in the functioning or expression of this receptor may underlie both a predisposition to and a manifestation of schizophrenia (Catts et al., Aus.& N.Z.J.Psych, 1997). Indeed mice genetically engineered to have a reduced level (5%) of NMDA receptor expression display social and sexual impairments in their interactions in addition to the stereotyped behaviours and increased motor activity of schizophrenia (Mohn et al., Cell, 1999), which could be ameliorated by antipsychotic drugs that antagonize both dopamine and serotonin receptors. In the social interaction test, an animal model of schizophrenia, PCP (a non-competitive NMDA antagonist) induces social isolation and stereotyped behaviour in rats which could also be overcome by anti-psychotic drugs which act at DA receptors (Sams-Dodd,Neurosci & Biobehav.Rev, 1998). Indeed, a measurable disturbance in glutamic and N-acetyl aspartic acid levels has been shown in unmedicated schizophrenics (Kishimoto et al., 1998).

The Acetylcholine Hypothesis

It is notable that the prevalence of smoking amongst schizophrenics is 3 times higher than in the general U.S. population (20-25%), as over 75% of schizophrenics smoke.

Further, nicotine withdrawal may temporarily worsen schizophrenic symptoms, suggesting that nicotine may help to control psychotic symptoms. The specificity of nicotine in its action upon a subtype of 'fast' acetylcholine receptors, known for their mediation of rapid excitatory signals, especially within the presynaptic terminal, infers a central role for acetylcholine in the aetiology of schizophrenia.

Cortical acetylcholine is known to mediate the detection, selection, and processing of stimuli and associations, and may additionally play a role in the filtering and allocation of other processing resources for these attentional functions. Attention is impaired if increases in cholinergic tone are blocked either by increasing GABAergic activity or by removing cholinergic inputs. Nicotine modulates both the failure to see inhibition of the P50 auditory-evoked response to repeated stimuli and the dysfunction in smooth pursuit eye movement associated with schizophrenia. Indeed D2 (and D1) antagonists attenuate increases in cortical ACh release stimulated by dopamine release within the Nucleus Accumbens, increases which are mediated by GABAergic neuronal projections to the basal forebrain that control the excitability of basal forebrain cholinergic neurons ([NA] DA Þ {-} GABA Þ {-} ACh [BF]). As DA release is elevated in the Nucleus Accumbens during both the acute phase of schizophrenia and as a result of the action of psychoactive drugs, it is suspected that cholinergic tone in the forebrain is altered in schizophrenia (Sarter et al., Annals N.Y.Acad.Sci., 1999). Further, this opens up the possibility of using nicotine as a treatment in schizophrenia, as it exhibits anxiolytic, attentional and cognitive benefits. Is schizophrenia attributable in part to a hypocholinergic deficit?

However, just as 'fast' nicotinic receptors may be implicated in schizophrenia, so may 'slow' acetylcholine receptors, inferred by the actions of clozapine which has a high affinity for these slow 'metabotropic' ACh receptors. This raises the possibility that the functional deficits in schizophrenia are attributable to changes in the function of slow as well as fast receptors.

The Serotonin hypothesis

The action of clozapine, an atypical antipsychotic used in the treatment of individuals resistant to dopamine antagonists which is noted to have a high affinity for serotonin receptors, suggested that serotonin (5-HT) may also play a role in the aetiology of schizophrenia.

Prozac (fluoxetine), an anti-depressant, as well as other serotonin reuptake blockers (SSRIs), are alleged to cause long-term deficits in memory, concentration and even mental disability, disrupting perceptions of reality and creating false memories (see Ann Blake Tracey, Prozac, Panacea or Pandora). Elevated levels of serotonin (5-HT) are found in schizophrenia, and SSRIs are alleged to have created an epidemic of suicide attempts (source: Prozac: Panacea or Pandora, by Ann Blake Tracey). Another potent serotonin-releasing agent, MDMA, otherwise known as ecstasy, causes hallucinations, memory deficits and other psychiatric symptoms affecting mood, cognition, and anxiety (McGuire, 2000) which are related to changes in serotonergic function.

The GABA hypothesis

Disputes have arisen as to whether GABA receptor function is altered in schizophrenics from evidence obtained from binding studies (Abi-Dargham et al., Neuropsychopharm., 1999), although GABAergic projections are certainly involved in mediating the effects of dopamine released from the nucleus accumbens upon activity in the prefrontal cortex.

Benzodiazepines, which act to increase the efficacy of transmission through GABA receptors and which are used both as sedatives and in the treatment of anxiety, have also been shown to ameliorate the core positive symptoms of schizophrenia (Pato et al., 1989). Indeed the brains of cocaine addicts, who exhibit schizophrenic symptoms, are more sensitive to benzodiazepines than those of drug-free individuals, indicating a change in GABA pathways in these individuals (Am. J. Psych., 1998). GABA receptors have further been shown to be important in the acquisition of behavioural sensitization to drugs that induce schizophrenia-like behaviours, such as methamphetamine.

Bogert's tempero-limbic hypothesis suggests two phases, a first in which a preferential loss of GABA receptors bearing NMDA glutamate receptors occurs, making the brain effectively hypofunctional for NMDA receptors, and a second stage in which the neural circuits altered by the loss of these GABAergic neurons are activated in late adolescence, but are consequently dysfunctional.

Recent interest has focused upon the reelin gene (RELN) whose expression is decreased by 50% in the telencepahlic GABAergic interneurons of the PFC, temporal cortex, hippocampus & also the glutaminergic cerebellar granule cells in schizophrenic patients. Reelin's signalling target, DAB1, is present in the neuroplasm of hippocampal & pre-frontal cortical pyramidal neurons as well as that of cerebellar Purkinje neurons (Impagnatiello et al., PNAS, 1998). Further, a second generation of telencephalic (including PFC, temporal cortex & hippocampus) region RELN is expressed in the adult cortex by horizontal and bitufted GABAergic interneurons, and thus RELN mediated signalling, similar to that which is operational during development, may continue during adult neurogenesis. Further, there are alterations in the level of GAD expression in GABAergic neurons in the PFC and changes in GABAA receptor density in the dentate gyrus and corticolimbic structures in schizophrenic patients (Impagnatiello et al., PNAS, 1998).

The anatomical and developmental theory of schizophrenia

Benes proposes that a developmental miswiring of dopaminergic inputs onto GABAergic neurons in the cortex occurs around birth (J.Psych.Res., 1997). Since the cortical dopamine system continues to mature until adolescence, the formation of misplaced connections during the normal ingrowth of dopaminergic fibres which occurs at this time, possibly exacerbated by stress, could trigger the onset of symptoms. Further evidence for a developmental onset comes from studies with rhesus monkeys irradiated with X-rays during foetal development which resulted in no ill effects until puberty, when schizophrenia-like symptoms such as poor working memory and hallucinations began to emerge (Castner et al., Soc.for Neuroscience, 1998). This lends further weight to the theory of Professor Goldman-Rakic that foetal brain damage predisposes an individual to the onset of schizophrenia at puberty.

Andreasen and colleagues (Biol.Psychiatry, 1999) have postulated that the diverse symptoms of schizophrenia are in fact attributable to a single disorder linked by a neurodevelopmental mechanism that results in the misconnection of neural circuitry, and specifically within the cortical-thalamic-cerebellar-cortical (CCTCC) circuit which is critical in the synchronous firing of neuronal centers involved in the smooth co-ordination of mental processes. Andreasen proposes that when the synchrony of the CCTCC circuit is impaired, the patient suffers from cognitive dysmetria, and this impairment of basic cognitive processes defines the hallmark of schizophrenia. Neuropathological changes occurring during the developmental stages of formation of the hippocampus has become a popular theory in predisposing an individual to schizophrenia, as the hippocampus is central in the processing, routing and filtering of sensory information which are known to be affected in schizophrenia (www.augsburg.edu/psych/vml/schizo.html). A loss of inhibitory neurons (GABA) appears to occur within the limbic lobe during development, concomitant with an infiltration of processes from excitatory neurons from elsewhere in the cortex, possibly predisposing to excitotoxicity. It seems that losses of neurons (gray matter) are reported to occur in schizophrenia by many groups, especially from the hippocampus, the amygdala, the superior temporal gyrus, parahippocampal gyrus and thalamus (www.augsburg.edu/psych/vml/schizo.html), and the PFC has been reported to decrease in volume in some patients. Indeed David Lewis suggests that a specific part of the circuitry of the PFC, specifically the chandelier neuron axon cartridge which controls information processing within other neurons in the PFC, is specifically reduced by 40% in schizophrenia (PNAS, April 1998). The density of expression of GABA transporters in these chandelier neurons is also decreased by 40%, again within these chandelier axon cartridges, an attractive and functional description of a neuron with divergent outputs (Impagnatiello et al., PNAS, 1998).

However, a significant proportion of individuals diagnosed as schizophrenic do not show symptoms until very late in life, seeming to argue against an exclusively developmental basis for schizophrenia (Owen & Castle, Drugs and Aging, 1999). Andreasen and others have also proposed that an enlargement of the fluid-filled ventricles of the brain may also be a feature of schizophrenia, although Staal and co-workers using computerized axial tomography demonstrated no correlation between the incidence of schizophrenia and ventricular enlargement, although it was predictive for the severity of symptoms (Schizophrenia Bulletin, 1999). Woods concluded that there is strong evidence AGAINST a classic neurodegenerative pathogenesis in schizophrenia, but that there is some support for prenatal developmental abnormalities and a loss of brain volume after the initial development of symptoms (Am.J.Psychiatry, 1998).

It may, however, be the case that the estimated 5% decrease in the volume of gray matter in the cortex may be due to a decrease in the number of connections between cells, including dendrites and chandelier axon cartridges, rather than primarily due to the loss of neurons or glia (Arch.Gen.Psych., 52, 1995). Thus, the poverty of thought, attention and memory associated with schizophrenia may reflect a poverty of brain cell interconnections, due to the excessive pruning of these axonal and dendritic connections, a process that appears to be at its height during adolescence (after puberty), although many now believe that the process begins before birth.

Steroids and the two hit model of schizophrenia

Steroid hormones influence a wide range of physiological functions from reproduction, stress, immune and inflammatory responses to behaviour, motor function and even cognitive performance (Rupprecht & Holsboer, 1999; Di Paolo, 1994). It has been proposed that there is a relationship between disrupted forebrain development and signalling by the steroid-like hormone retinoic acid, which is produced by a developmental layer of cells derived from neural crest known as the mesenchyme (within the anterior neural tube), responsible for the induction and differentiation of adjacent epithelia. It is this induction mediated via interaction between the retinoic acid-producing mesenchyme and the anterior surface epithelium of the embryo that guides differentiation and PATHWAY FORMATION. It is thought that such a developmental flaw, either inherited or environmental, may constitute the "first hit" in the genesis of schizophrenia.

Schizophrenia is extremely uncommon before adolescence and puberty, suggesting that the surge of altered steroid hormone biosynthesis associated with this stage of development, and possibly also those steroid hormones which are elevated in response to stress during these critical career-forming years of life, may also be implicated in the aetiology of schizophrenia. Further, sex hormones have been shown to influence dopaminergic activity (Di Paolo, 1994). Such surges in hormone levels may constitute the "second hit" in schizophrenia, facilitating excitotoxicity or oxygen radical formation that leads to neuronal damage. Indeed, stress is well known as a common precursor of the first episode of psychosis.

There is further evidence that altered levels of sex steroid hormones are associated with schizophrenia. Male schizophrenics were found to have higher levels of Lutenising Hormone (LH) and testosterone than healthy subjects, presenting a puberty-like profile, and female schizophrenics higher levels of LH and lower levels of oestrogen, in effect a menopause-like profile (Kulkarni et al., Schizophrenia Res., 1996).

Schizophrenia: An inherited disordering of the mind?

Interest in an inherited causality for schizophrenia came from observations that schizophrenia often appeared to be clustered in families, an identical (monozygous) twin having a 40 to 50% chance of developing the illness, and a child of a schizophrenic parent around 10%. These rates are statistically high, but suggest neither classic Mendelian patterns of inheritance, nor exclude the effects of family or local environmental influences. A study of relatives of adoptees, in an attempt to minimize environmental influences, also found that schizophrenia is concentrated within biological families, the incidence of schizophrenia being slightly, but significantly elevated (5.1%) within the biological relatives of adopted schizophrenics (Kety & Ingraham, J.Psych.Res, 1992). Psychiatric studies have suggested that schizophrenia is a disorder with multifactorial inheritance, i.e. involving many genes present in many different locations throughout the bank of human 'chromosomal' information, as schizophrenia does not follow simple Mendelian inheritance.

A disruption of genes that govern the neural crest-mediated, RA-dependent induction and differentiation in the forebrain such as Pax-6 & Gli-3 might be implicated in schizophrenia (LaMantia, Biol.Psych., 1999) and perhaps crucially, the secreted extracellular matrix protein reelin (RELN) may play a role in schizophrenia. Linkage analysis of microsatellite regions in families with patterns of schizophrenia have revealed that genes predisposing susceptibility to schizophrenia may be present on chromosomes 1q, 5q, 6p, 8p, 13q, 15q, 18p and 22q (Shastry, Neurogenetics, 1999), one of which (15q, locus 14), has been associated with the a7 nicotinic receptor underlying the P50 deficit of schizophrenia (Adler et al., Biol.Psych.1999) . Regions on chromosomes 3, 9 and 20 have also been proposed to be candidates for schizophrenia genes (www.augsburg.edu/psych/vml/schizo.html). Thus not only may the many hundreds of genes involved in neuronal signalling and the development of the nervous system be potential factors in schizophrenia, but at least 8 chromosomal regions, each of which potentially may contain many genes which predispose to schizophrenia, have been variably associated with linkage analysis. This observation rather supports Andreasen's contention that any "misconnection of neural circuitry ... within the cortical-thalamic-cerebellar-cortical circuit which is ... involved in the smooth co-ordination of mental processes... when .... impaired, (causes) the patient (to) suffer from cognitive dysmetria". This variation in causality is further suggested by the observation that 2% of schizophrenics have microdeletions on chromosome 22(q11) which may be associated with altered speech and learning difficulties (Weinberger, Biol.Psych., 1999). In addition, variations in the D5 receptor gene have been found in some (but not all) schizophrenics (Feng et al., Am.J.Med.Gen. 1998). Schizophrenia may result from small so-called unstable expansions of repeating trinucleotide (CAG)n and (CGG)n 'triplet' microsatellite sequences associated with dominant genes associated with ataxia (a disorder of movement) such as Spinal Cerebellar Ataxia type 1 (SCA1, Fischer, Med.Hypoth. 1998). Intriguingly, such an expansion of a random repetition of the nucleotides at the beginning of a gene that encodes for a K+ channel (hKCa3) in one of the chromosome regions believed to be associated with schizophrenia, may be involved in the regulation (inactivation) of NMDA receptors (Li et al., BBRC, 1998).

Clearly the evolution of parallel processing within the neocortex in higher mammals, which requires a prolonged period of development *ex utero* before language and social function are fully developed (perhaps as long as seventeen years after birth in humans), leaves many potential avenues for disruption or injury, some of which we may refer to as schizophrenia. Alternatively, are schizophrenics an example of a previously advantageous polymorphism of form or function that is currently disadvantaged in the high information throughput, technocratic and intensively socially interactive society of today. For example, genes that protect against typhoid and malaria in 'single dose' are lethal in double dose in cystic fibrosis and sickle cell anaemia respectively. Would an absence of sensory filtering have been advantageous to a Neolithic hunter whose auditory and visual responses remained uninhibited to sounds and sights of predator or prey? Is schizophrenia an example of the 'Odyssian personality' which gives rise to an advantage in evolutionary adaptation to changing times (Rison, 1998)?

Other prevailing theories on the causation of schizophrenia

Five studies have suggested that being born or raised in an urban area is a risk factor for schizophrenia. E.Fuller Torrey and Yolken have proposed that an infectious agent transmitted through household crowding maybe responsible (Schizophrenia Bulletin, 1998). Even neurotrophic viruses have been suggested. Royce Waltrip believes that the Borna disease virus, an agent that was thought only to affect horse and sheep, causing brain inflammation through an immune response, is also a causative agent in schizophrenia. Waltrip and co-workers found antibodies to the Borna virus in 9 of 25 schizophrenic twins in a cohort NIMH study. Previous investigators have suggested that a first "hit" risk factor for schizophrenia may be viral infection *in utero*, although studies in Britain (Cannon et al., Br.J.Psychiatry, 1996) and the Netherlands (Takei et al., J.Psych.Res, 1995) indicate that exposure to the influenza virus during gestation poses no substantial risk for schizophrenia.

Other factors predisposing to schizophrenia include perinatal hypoxia, autoimmunity (Jones & Cannon, 1998). Dohan's Hypothesis proposes that Schizophrenia is an inherited predisposition which interacts with an overload of dietary proteins such as casein, glutens or gliadins. In a potentially related finding, exorphins, morphine-like compounds produced from milk protein which are taken up into the brain, are elevated in 95% of autistic and schizophrenic children, especially b-casomorphin-7. In contrast, Peet & Puri have proposed that schizophrenia is caused by a depletion of certain fatty acids in the membranes of nerve cells, symptoms which are ameliorated with dietary intake of EPA.

Other neurotransmitter systems, not all of which have been identified to date, may be involved. For example, norepinephrine (NE) is proposed to a role in the induction of psychosis and is further evidenced by the observation that a prenatal exposure of rats to amphetamines causes an increase in NE levels in the PFC. These observations are taken to be suggestive of a hyperactive NE system, resulting in psychotic behaviours (Nasif et al., Brain Res., 1999).

Other causative hypotheses take into account socio-developmental factors. Dysfunctional relations with, or absence of mother were thought by some to be *schizophrenigenic*, although this argument is now widely refuted. Social stressors in urban settings are thought to facilitate the onset of disease in vulnerable persons (www.ndmda.org/schiz.htm). Indeed the loss of neuronal and/or connective tissues in schizophrenic patients due to the neurotoxic effects of overtransmission of dopamine or excitatory amino acids (glutamate; www.augsburg.edu/psych/vml/schizo.html) may be attributable, in part if not in whole, to stress.

In conclusion, Andreasen has demonstrated that there is connectivity between nodes (or clusters of neurons involved in processing) in the pre-frontal cortex, the thalamic nuclei and the cerebellum. Any disruption in this circuit may result in a cognitive impairment or dysmetria, leading to a difficulty in prioritizing, processing, coordinating and responding to information:- central deficits in schizophrenia. Specific neurochemical disruption involving the release or reception of the neurotransmitters serotonin, glutamate, dopamine, acetylcholine, norepinephrine or GABA may be sufficient to wholly or partially mimic the symptoms of schizophrenia. Thus schizophrenia may be regarded as a heterogeneous dysfunction of cognitive and sensory processing.

Schizophrenia: A contemporary epidemic?

Schizophrenia is a disabling condition with an age of onset which is earlier for men (15-25) than for women (25-35), with a lifetime prevalence of 1.3% within the U.S. population (NIMH, or 2-3 million Americans, of whom fewer than 1 in 5 recover fully. About half of all schizophrenics will attempt suicide at least once, 10-15% of whom will be successful. Schizophrenia has an attributed, estimated annual direct and indirect cost to U.S. of $30-48 bn, and there are related costs due to the high prevalence of substance abuse and cigarette smoking amongst schizophrenics. More intriguing and indicative for the possible role of environmental factors in the aetiology of schizophrenia, is the epidemiology of the disease. Specifically, the question is whether the incidence of schizophrenia is rising (or falling) as a consequence of changes in lifestyle and work practices? Surprisingly there is little evidence available in the literature to reveal any such patterns, although one paper was found which predicted an incipient epidemic of schizophrenia amongst young black males, a high-risk group within the U.S. population (Turns, Ann.Med.Psychol., 1980).

Who constitute high-risk groups for schizophrenia?

Low socio-economic status has NOT consistently been shown to be a risk factor for schizophrenia although this remains contentious. Several high-risk groups have been identified, however, in addition to late adolescent males, especially blacks (Turns, 1980), and women during menopause, childbirth, and pregnancy. Against a background prevalence of 1.3% for the general U.S. population (and approximately 1% in the U.K.), a recent follow-up study of mental health amongst non-German speaking immigrants indicated that 38.7% were schizophrenic, whilst only 8.3% were diagnosed as impaired at the time of entry, although this was admittedly seen as an overestimate due to language difficulties (Haasen et al., 1998).

Studies amongst the homeless in Germany have indicated that schizophrenia is significantly over-represented amongst both women (21.9%, Greifenhagen & Fichter, 1997) and men (12.4%, Fichter et al., 1997). These studies were the lowest estimates available, and likely an underestimate as some studies found that between a third and two-thirds of the homeless population were afflicted by schizophrenia. Intriguingly, whilst the rates for effective psychoses were low among men in Greenland, a possible model for 'social' isolation in comparison to the Danish mainland, rates for schizophrenia and suicide were found to be very high (Lynge et al., 1999).

Do psychoactive drugs *cause* or merely mimic schizophrenia?

Although it is widely accepted that whilst drugs which act by 'blocking' NMDA receptors (e.g. PCP, ketamine), dopamine transporters (DAT; cocaine/methamphetamine) and serotonin transporters (Prozac, MDMA) mimic in part or in whole the symptoms of schizophrenia, it is 'said' that they do not cause schizophrenia as their actions are transient, lasting only for the duration of drug action, cocaine, for example, having a half-life of only some 6 minutes. However, this presupposes that there is neither a progressive augmentation in their action (behavioural sensitization) which may eventually become sustained in the absence of drug and secondly that the drugs themselves do not cause functional neurotoxic damage at these self-same sites of action which are sufficient to induce the 'schizophrenia-like' symptoms.

However behavioural sensitization is seen in both humans and animals exposed to methamphetamine, especially in regard to hallucinations, suggesting a prolonged change in the neurochemical balance within the brain (Ito, Hokkaido J.Med., 1999). Amphetamine especially induces psychosis after prolonged, frequent high dose exposure. Further, drugs which increase dopamine levels by blocking its reuptake by the DA transporter into nerve terminals, such as cocaine and amphetamine, cause positive symptoms of schizophrenia such as increased locomotion, hallucination and other psychotic states at high concentration. A clear correlation between the degree to which cocaine blocks this transporter and a cocaine abuser's euphoric feelings has been demonstrated, at least half of these transporters having to be blocked in order for the user to perceive cocaine's euphoric effects (Volkow, Nature, 1998). Ritalin (methylphenidate), a stimulant sharing a molecular mechanism of action with cocaine at the level of the DAT transporter, was administered to both cocaine addicts and non-users in PET studies. The cocaine (ab)users showed reduced dopamine responses to Ritalin in the striatum, a region linked to motivation and reward, and an abnormal increase in the DA response in the thalamus associated with intense experiences of cocaine cravings in addicts, not observed in the control group. Repeated exposure to cocaine causes long-term changes in behaviour ranging from addiction to behavioural sensitization which are related to the nigrostriatal system of the basal ganglia, of which the striatum is a part. Chronic cocaine application causes alterations in the inducibility of bZIP transcription factors, immediate early gene expression (cFos) and changes in the expression of ensembles of striatal neurons which express these proteins, suggesting network level adaptations and a functional reorganization of these basal ganglia circuits (Moratalla et al., Neuron, 1996).

Cocaine, which is both a local anesthetic and vasoconstrictor, reaches the brain within minutes and produces a fast acting and short-lived period of euphoria mediated by an elevation of DA levels followed by a dysphoric 'crash' with depression, anxiety, craving and fatigue. Repeated doses of cocaine lead to the constriction of blood vessels in the brain and may at high doses cause brain hemorrhage, heart failure or stroke. There are more than an estimated 2 million cocaine addicts in the U.S. and 3.8 million users (NHSDA www.usdoj.gov/dea/concern/cocaine.htm). Perhaps more concerning, 4.7 million Americans have tried methamphetamine in their lifetime (otherwise known as *speed, ice or crystal*). Amphetamine was first marketed as Benzedrine in the 1930s when it was used to treat ADHD, and during WWII methamphetamine (Methedrine) and dextroamphetamine (Dexedrine) were used to motivate troops. Methamphetamine, or **meth**, is preferred by many drug-users as it produces a more gradual and sustained euphoric effect (a 'cool smoke') than crack cocaine (which results in a 'hot smoke'). Further, it is cheaper to purchase as it is produced in bulk by over 1,600 reported methamphetamine labs throughout Mexico and the U.S. (DEA figures, www.usdoj.gov/dea/concern/meth.htm). Use must rise with supply, as supply rises to meet demand, and there has been an exponential rate of increase in 'meth' lab seizures throughout the 90s. Does this not suggest an epidemic of schizophrenia-like symptoms?

Chronic use of methamphetamine produces a psychosis that resembles schizophrenia, characterized by paranoia, picking at skin, preoccupation with one's own thoughts and hallucinations. Disorganized and violent behaviour is often seen amongst chronic abusers. The long-term changes seen in methamphetamine users may have a molecular basis, and indeed drugs such as PCP and cocaine induce lasting changes in the patterns of gene expression which may last for weeks following drug withdrawal. For example, a single dose of PCP alters the neurochemistry of the anterior cingulate cortex and changes patterns of c-Fos expression (an immediate early transcription factor) in the striatum (Turgeon & Roche, Neuroscience, 1999). Repeated amphetamine exposure has been shown to alter DA systems and to induce behaviours reminiscent of both the positive and negative symptoms of schizophrenia in primates. Indeed, behavioural sensitization to amphetamines was present after 5 days of chronic exposure and persisted for as long as 28 months after withdrawal (Castner & Goldman-Rakic, Neuropsychopharm., 1999).

Monkeys treated with PCP twice daily for 2 weeks displayed sustained deficits in tasks requiring PFC function. This repeated exposure to PCP caused a reduction in both the basal and evoked utlilization of DA in the dorsolateral PFC, consistent with observations in schizophrenic patients (Jentsch et al., Science, 1997).

Do psychoactive drugs cause brain damage?

It is well established that repeated treatment with psychostimulant drugs produces changes in brain and behaviour that far outlast their initial neuropharmacological actions. These changes contribute to the development of dependence and of addiction and psychosis. A range of tests on *chronic abstinent* cocaine users and 'drug-free' controls revealed that cocaine caused reduced executive functioning, visuoperception, psychomotor speed and manual dexterity (Bolla et al., J.Neuropsych. Clin. Neurosci, 1999). Another study showed that cocaine abusers had deficits in attention, concentration, new learning, visual and verbal memory, word production and visuomotor integration (Strickland et al., J.Neuropscyh.Clin.Neurosci., 1993), consistent with persistent decrements of cognitive function, as occur in schizophrenia. Chronic cocaine abuse may be a neuropsychiatric syndrome, causing changes which include EEG abnormalities, seizures and decrements in neurobehavioural performance, in addition to the acute psychotic and paranoid states associated with administration and the depression associated with withdrawal. Cadet and Bolla (Synapse, 1996) propose that chronic cocaine use is an example of a 'disconnection syndrome'. Further, the vasoconstriction that cocaine causes leads to sustained hypoperfusion and ischaemic damage in brain regions, leading to significant cerebral hypoperfusion in the cortex (Strickland et al., J.Neuropscyh.Clin.Neurosci., 1993).

Which brain regions are implicated in schizophrenia & drug neurotoxicity?

Volkow's group in 1998 showed that cocaine addicts have evidence of persistent damage to their brain chemistry, their brains being more susceptible both to seizures and sleep abnormalities. Volkow's group showed that the brains of cocaine addicts were more sensitive to drugs which enhance GABAergic activity because of damage to these GABA pathways which transmit pleasure signals to other regions of the brain.

Whilst neither amphetamine or cocaine have been shown to cause neurotoxicity in the nucleus accumbens (Xu et al., Brain Res., 2000), mice engineered to express higher levels of DAT transporters (THDAT mice), whilst habituating more rapidly to novel environments and displaying an enhanced reward in response to cocaine, show 50% greater losses in dopaminergic neurons in response to MPTP toxin treatment than controls, indicating that dopaminergic neurodegeneration occurs in response to substance abuse (Donovan et al., 1999). The appearance of decreased N-acetyl compounds in the frontal cortex of **abstinent** cocaine abusers is further indicative of sustained neuronal injury in this region (Chang et al, Am.J.Psych., 1999), and even subchronic cocaine administration for 5 days results in a pronounced degeneration in the lateral habenula (LHB) and its primary efferent tract (output) the *fasciculus retroflexus* (Ellison, 1992).

Immunostaining revealed a marked decrease in the density of GABAergic, but not glutaminergic, nerve terminals in the lateral habenula (LHB), but **not** in the nucleus accumbens, suggesting an exquisite sensitivity of this specific neuronal population. It is suggested that a decrease in inhibitory GABA activity leads to increased excitatory transmission through LHB glutaminergic neurons and a resulting neurotoxicity within and consequent degeneration of the fasciculus retroflexus (Meshul et al., Synapse, 1998). The habenula is the chief relay nucleus of the descending dorsal diencephalic (DDD) system which is an important link between limbic and striatal forebrain and lower diencephalic/mesencephalic centers. The DDD system has functional connections by which to modulate sensory GATING through the thalamus (sensory), pain gating through the central gray and raphe (pain), and motor stereotypies and reward mechanisms through the substantia nigra and VTA. Lesions to the habenula alter a variety of behaviours, and damage to this habenula pathway constitutes an excellent candidate for producing behaviours which occur during psychosis (Ellison, Brain Res.Revs., 1994). At recreational doses, methamphetamine causes a decrease in striatal DAT transporter density in primates, indicative of a loss of DA neuron axonal terminals, which were further depleted at higher doses (Villemagne et al., J.Neurosci, 1998). Further, continuous amphetamine (but not cocaine) administration is neurotoxic to DA innervations in the caudate nucleus (Ellison, Brain Res.Rev., 1994).

Ecstasy, or 3,4 methylenedioxymethamphetamine (MDMA), evokes both a 'calcium-independent' release of brain monoamines (*i.e. serotonin, dopamine, norepinephrine*, Johnson et al., 1986) and inhibits their inactivation by reuptake into nerve terminals (Steele et al., 1987). At sufficient concentration, MDMA may cause a depletion of serotonin (5-HT) transporters ranging from 44% in the pons to 89% in the occipital cortex, indicative of widespread damage to serotoninergic neurons in these regions (Scheffel et al., Synapse, 1998). Whilst some regions recovered after 9 months e.g. hypothalamus, others areas such as the neocortex did not appear to recover from what appears to be a loss of serotoninergic axonal terminals (neuronal outputs).

Even though no apparent neurotoxicity is apparent in the nucleus accumbens, repeated application of cocaine or amphetamine causes alterations in the morphology of dendrites and dendritic spines, increasing both dendritic branching and spine density (synaptic contacts), thus reorganizing patterns of synaptic connectivity in the nucleus accumbens and prefrontal cortex (Robinson & Kolb, Eur.J.Neurosci., 1999). In addition, cocaine inhibits neuronal differentiation in PC12 cells in response to Nerve Growth Factor (NGF, Zachor et al., Mol.Gen.Metab., 1998), and cocaine further causes programmed cell death (apoptosis) in embryonic neuronal precursor cells, which may be suggestive of a more general process of reorganization of plasticity and connectivity.

Telling insights from animal models of schizophrenia

"The absence of an animal model that accurately approximates schizophrenia limits current research into the pathophysiology of this disorder" (O'Donnell & Grace, 1999)

Normally the presentation of a weak stimulus immediately before a stimulus which is sufficient to startle a rat will lead to a decrease in the magnitude of the resultant startle response, a phenomenon termed **prepulse inhibition (PPI)**. This provides a measurement of sensorimotor gating which is also deficient in schizophrenia patients. Intriguingly isolating rats from weaning to adulthood also causes this deficiency in the PPI response and hyperactivity, without apparent evidence for a genetic predisposition, as it occurred regardless of the breeding strain used. However, the PPI deficits induced by isolation only emerged during or after puberty, in contrast to the hyperactivity

(Bakshi & Geyer, Physiology, and Behaviour, 1999). Moreover, these effects persisted after the cessation of isolation (Domeney & Feldon, Pharm, Biochem & Behav., 1998). Even a short stressful life event, such as a 24-hour maternal deprivation, was sufficient to introduce the reduction in the PPI of the startle reflex observed in schizophrenics. However, this effect was not seen until puberty, suggesting that the action of stress steroid hormones may combine with those of sex steroid hormones to alter cortical development at this time (Ellenbrook et al., Schizophrenia Research, 1998). Thus even early life events may have a profound influence on information processing and social functioning. Perhaps pertinent is the observation that schizophrenics accommodated within the community rather than within the asylum suffer more relapses in a high 'expressed emotion' environment wherein they receive negative criticism, overinvolvement by certain members of the family, expressed hostility, and prolonged contact with such individuals (Green, www.priory.com/schizo.htm).

The hippocampus has been implicated as central in the formation of memory and learned behaviours and in particular spatial memory for places and resources. Rats with neonatal lesions in their ventral hippocampus, but not medial PFC (Lipska et al., 1998), showed less time spent in interaction and an enhanced level of aggression, which was not due to anxiety. This effect however occurred only in young rats and not in those lesioned after weaning, indicating that damage to cortical integrative circuits sustained before or during the period of learned social behaviours can result in schizophrenia-like 'asociality' (Becker et al., Psychopharmacology, 1999). The causation of this damage need only extend as far as the expression of a given gene, as mice expressing only 5% of normal levels of an NMDA receptor subunit show the impaired social and sexual functioning and stereotyped behaviours believed to be related to schizophrenia. Introduction of PCP induces stereotyped behaviour and social withdrawal in rats, the animal behavioural correlate of schizophrenia, which was similarly alleviated by DA anti-psychotics (Sams-Dodd, Rev.Neurosci., 1999).

The slow decline in condition (prodromal phase) preceding of schizophrenia lasts several years (Hafner et al., Acta Psych.Scand., 1999) and social disability emerges 2-4 years prior to psychiatric admission. It has yet to be proven beyond doubt that the cognitive decline, deficit in PPI and other deficits present in schizophrenia occurs prior to, concomitant with, or as a result of social decline and isolation.

An alternative theory for the causation of schizophrenia:

The four 'S"s: Stress, steroids, solitude and sensory deprivation

Stress & Steroids

It is held that stress precipitates the positive psychotic symptoms of schizophrenia in many sufferers, and stress has been noted to precipitate positive symptoms in schizoid personality disorder sufferers who normally display only the negative symptoms of schizophrenia (Sverdlov, 1998). Indeed stress is a common precursor of the first episode of psychosis following a long prodromal phase where only negative symptoms are seen.

Stress induces the release of neuroactive steroids such as corticosterone from the hypothalamic-pituitary-adrenal axis, which like sex steroids including progesterone, have profound modulatory actions upon the function and expression of receptors and processes involved in the nervous transmission and processing of information. For example Cho and Little (Neuroscience, 1999) showed that perfusing slices from the ventral tegmental area, an area implicated in drug dependence and reward, with the steroid hormone corticosterone, released in response to stress, caused an increase in sensitivity of the activity of dopamine-releasing neurons to the three 'fast' glutamate receptor subtypes AMPA, kainate and NMDA.

These pacemaker neurons release DA into the prefrontal cortex and nucleus accumbens as part of the neural mechanisms of reward and pleasure, and this increased sensitivity to glutamate is observed within 15 minutes of application of corticosterone, showing that stress can rapidly modulate neuronal activity and make neurons exquisitely sensitive to glutamate-mediated excitotoxicity. Indeed Gardner and others have proposed that mammals seek to maintain elevated DA levels within the Nucleus Accumbens, released from projections arising within the VTA, as a reinforcement of a positive activity, behaviour or environment. Stress is catabolic, and pleasure, resulting in an elevation DA levels in the Nucleus Accumbens, PFC and hypothalamus is anabolic. So sex, food and social reward (for achievement or as a result of desirable or gainful behaviour) can be argued to represent 'trophically' advantageous behaviours, and their reinforcement is mediated, at least in part, by enhanced activity within the dopaminergic system. However, this pleasure-seeking or reward behaviour may be mimicked or bypassed by the use of psychoactive compounds which mimic or 'short-circuit' this enhancement of DA release, or else by abuses of food, video games or sexual activities, features of the densely-populated and technologically 'enriched' modern city environment.

An increase in the functional weighting of excitation through glutamate receptors over inhibition through GABA and glycine receptors leads to an increase, both directly and indirectly, in levels of intracellular calcium, the universal signal of intracellular excitability. Excess calcium entry into the cell, which may occur due to cerebral ischaemia (loss of oxygen), head injury, excitotoxic drugs or even excessive nerve cell activity, causes damage to nerve cells and their connections (terminals), and at very high levels can even cause their death. Hence extreme stress as a result of traumatic life events or as a process of active social isolation, either alone or in combination with psychoactive drugs and possibly excessive excitatory transmission, may lead to the damage of intricate neuronal circuits and hence of cognitive processing and behaviour.

Excitotoxicity may in part explain the cognitive and functional decline observed in schizophrenia, but may not be the only possible explanation for a decline in the degree of neuronal arborization (dendrites and axons) which have been suggested to occur in schizophrenia, and accordingly further experimental evidence must be afforded in support of this model.

Solitude & Sensory deprivation: the activity dependence of survival

Neurons extend processes during development and repair which are guided by factors to their programmed targets by molecules secreted from their target cells known as neurotrophins. However, the establishment of functional neuronal connections requires the secretion of signals from the outgrowing axon terminals. Indeed this process has been shown to occur even in the absence of synaptic signalling, as mice that that lack the synaptic protein *munc 18-1*, essential for the release of neurotransmitters, nonetheless form 'normal' layered structures, fibre pathways and morphologically defined synapses during development (Verhage et al., Science, 2000). However, in the absence of functional synaptic transmission, the neurons in this 'knockout' mouse undergo programmed cell death (apoptosis), leading to widespread neurodegeneration. Thus neurons require activity (neurotransmission) to survive, and if such activity is absent they will die as part of a pruning process that maintains only those neuronal connections which are useful, i.e. functional, perhaps only a fraction of the number that are originally formed. Thus there exists a dual influence of neuron upon target, and of target upon neuron, by the release of neurotrophic factors such as Nerve Growth Factor (NGF). Thus there is a both a genetically-programmed and use-dependent refinement of connections in the nervous system during development, and the more sophisticated cortical circuits associated with information processing, language and higher social functioning which emerge both later in evolution and during development, require functional validation for the establishment of their final patterns of connectivity (Shatz, 1997).

Dr. Fred Gage and co-workers have shown that new cell growth continues to occur in the hippocampus in patients aged from 55 to 70, and in fact, new cells are constantly being generated within the hippocampus and neocortex of adult monkeys (Gould et al., 1999). Further it has been shown that associative learning enhances adult neurogenesis in the hippocampus in rats (Gould et al., 1999), whilst stress, via the action of glucocorticoid steroid hormones, in contrast inhibits neurogenesis in the hippocampus and causes an increase of potentially neurotoxic glutamate release within the hippocampus (Gould & Tanapat, 1999).

Brain cells in most people do not, in fact, die when we age (excepting senile dementia, Alzheimer's etc..), rather the number of connections that they form diminishes, and this connectivity is believed to correlate with a neuron's "computational power", and is enhanced in response to growth factors called neurotrophins which are released when neurons are stimulated (McAllister et al., 1996), for example by new learning. Contextual learning in the hippocampus has been demonstrated to evoke the release of neurotrophin (BDNF), thereby increasing the extent of arborization of dendrites and thereby the potential computational power of the circuits involved (McAllister et al.,

1996). Conversely, removing neurotrophins causes dendrites to atrophy, which suggests that a lack of brain activity causes a loss of computational activity and mental decline (Katz, McAllister & Lo). In fact, neurotrophins help to maintain our intelligence and mental function via their release from active neurons, although it might be noted that neurons may be activated by a range of stimuli from sound through thought to smell and emotion. Thus, sensory and social deprivation may cause a loss of computational power in regions of the brain involved in sensory processing, language, and higher social intelligence, with the attendant consequences of social decline and lost opportunity.

Such is the degree of plasticity present in the nervous system for the growth and regeneration of new cells and processes, that even after the loss of sight or even actual regions of the brain, function can be restored or replaced in whole or in part by other regions in a time and **use**-dependent manner. For example, monkeys who lose part of their brains involved in the control of fine finger touch and manipulation recover this function in neighbouring areas over time with **use** in areas that previously had no such attributed function (Xerri et al., 1998). Similarly, people who lose their sight after development show a shift in functional activation of their visual cortex from sight towards somatosensory touch upon learning Braille.

As stated before, it may be the case that the interpretations of an estimated 5% decrease in the volume of gray matter in the cortex observed in schizophrenia may be attributable to a decrease in the number of connections between cells, such as the dendrite and chandelier axon cartridge, rather than primarily a loss of neurons or glia per se (Arch.Gen.Psych., 52, 1995). The poverty of thought, attention, and memory associated with schizophrenia may reflect an acquired and progressive impoverishment of brain cell interconnections, due to an excessive loss or pruning of axonal and dendritic connections.

In search of an explanation for schizophrenia: Mice, men or the mind?

Each model of study the mind, the brain and the animal model affords advantages and disadvantages, in effect forcing a pronged approach to the study of multifactorial traits such as schizophrenia or depression.

Advantages	Disadvantages
Mice allow genetic manipulation and the rapid expression of traits	The human **mind** does not allow an easy examination of the underlying molecular, genetic or biochemical causations of mental 'illness'
'Men' allow linkage analysis and DNA testing of defined psychological profiles determined by psychometric parameters	**Men** are a heterogeneous and poorly controlled mixture of environment, experience and inheritance
The **'mind'** allows the study of the dynamic functioning of the brain in response to fixed stimuli, measured by imaging or electrically evoked potentials	**Mice** do not sell stocks and shares, pass stressful examinations or manage their finances, but they do behave...socially and trophically

A transgenic solution?

Much insight into the mechanisms of neural signalling has been gained from transgenic animals, i.e. those which have been genetically altered to change the amount or sequence of a given gene is expressed. For example, mice that have been genetically altered to express only 5% of an NMDA glutamate receptor gene (NR1) exhibit the social and sexual impairments, stereotyped behaviours and increased motor activity typical of schizophrenia (Mohn et al., 1999). Perhaps but fortune rather than deliberation, knock-out mice lacking a gene that synthesizes neuronal nitric oxide (NO, nitric oxide synthase, NOS), an intended animal model for stroke, became socially aggressive and lacked normal patterns of social interaction, forcibly mating with unreceptive females and killing cage mates.

The problem with transgenics is that any mutation might conceivably affect development as well as function in the adult animal, and these are often difficult to distinguish. A further dilemma is that mental 'illness' is already known to be an example of multi-factorial inheritance, and a plethora of potential genes may likely be involved in schizophrenia and related disorders and that these alterations may be very subtle and inter-linked, making 'black-and-white' knockouts in mice a debatable model for mental illness, regardless of their scientific usefulness.

Gene expression

A great number of gene products have been implicated in schizophrenia, which may be modulated at many levels, from mutations that affect their final form or their level of expression to genetic and epigenetic factors such as satellite repeats, chromosomal recombinations, acquired gene silencing through methylation, and aneuploidy, an altered number of chromosomes in a given cell. These are all changes at the fundamental DNA level of encoding and exclude variations in functional gene expression at the level of expression or processing of the genetic transcript (RNA) or its final product, the protein. Even though we can now more finely control expression, the sheer array of possible targets and the variety that may stem from changes in RNA or protein processing possibly from an apparently unrelated gene means that it may be easier to work backward...

At what levels do we study genes, and by what do we measure when we look at genes at different levels and do we lose any information in the process? The completion of the human genome project will give us a map of the location and the sequence of all the genes in the human genome. This will allow us to study variety in human polymorphisms, chromosomal rearrangements, and mutations.

This, however, does not tell us how disease or change may arise by a variability in the expression of the gene. For that we have to go to the basic message and study the expression and sequence of the RNA that must be first synthesized before protein can be made.

The importance of environment to development and function

The prevailing research provocatively suggests that an enriched social environment is necessary for the development of 'normal' behaviour and enhanced cortical thickening, as rats isolated after weaning develop behaviours reminiscent of schizophrenia and mice bred in enriched environments show both an increased cortical thickening and a greater number of hippocampal neurons (Kemperman et al., 1997). Indeed Romanian orphans raised in an isolated and impoverished environment showed a level of development that was so markedly retarded as to raise the question as to whether schizophrenia with its attendant loss of cognitive function is not a mild manifestation of an activity-dependent developmental disorder caused by social and sensory deprivation.

Why should social contact be so important in cognitive performance, decision-making and the processing of sensory stimuli? The answer may be suggested in the formation of ocular dominance columns in the visual cortex which is itself dependent upon neural activity which is visually driven (Hubel and Wiesel, 1998). By depriving one eye of visual information for several weeks during an early critical period of development (monocular deprivation) there was a marked change in the patterns of activation in the primary visual cortex. Cells in layer IV of the visual cortex were now activated only by input from the eye that had remained open, even if the other eye was still functional in the detection and transmission of light signals. In other words, a constant sensory input from both eyes is required for the correct development of binocular processing of light information and the attendant orderly patterning of brain cells in the visual cortex. Anatomically, the terminal arbors of the axons of the lateral geniculate nucleus which were supplied by the uncovered eye were considerably more extensive than those supplied by the deprived eye. As an explanation for this phenomenon, it has been found that there is an activity-dependent release of neurotrophic factors by cortical neurons and that this may affect the pattern and extent of wiring in the visual cortex. In fact, this competitively-based loss of function from the covered eye could be specifically overcome by localized infusion with the neurotrophin NT-4 (Riddle et al., 1995).

In plasticity there is hope. If sight is lost after its development the areas of the cortex attributed to vision show considerable plasticity in that they remodel from being activated primarily by visual information, to being activated by another sensory input, somatosensory touch, an acquired plasticity, (learned) from the acquisition of Braille reading skills. Similarly, social engagement and other high-level forms of play and linguistic interaction, which constitute 'higher' primate behaviours and which are associated with the evolutionary enlargement of the forebrain and association cortex, must not only be acquired through development, but must be used or else they may diminish. Neurons form connections in response to stimulation, at least in part induced by the activity-dependent release of neurotrophins, and if these 'activity-dependent' neuronal wiring patterns are not induced by sensory input they may not form (Romanian orphans), and if they are not maintained they may diminish or be 'functionally' lost (schizophrenia?). Another possibility is that stress or information 'overloading' (excessive activity) may cause damage to the neuronal processes of these cortical and subcortical structures resulting in cognitive disruption and a loss of social functioning, concentration, disrupted speech etc.

Breeding a natural animal model for schizophrenia .... (and deciphering the changes)

"...the genetic dissection of quantitative behavioural traits, such as mood, personality and intelligence... pose new problems for gene cloning experiments...one way forward is by using animal models...and an efficient strategy for detecting sequences that give rise to quantitative behavioural traits can be devised in the mouse" (Flint & Corley, 1996).

There are almost certainly changes in the patterns of genes expressed underlying mental disorders such as depression or schizophrenia. It has been known by generations of animal breeders that specific behavioural characteristics may be bred into domestic animals such as dogs, sheep or horses. The same strategy has been used, for example, to breed rats that are chronically depressed in their responses to painful stimuli, or mice that are unusually aggressive (e.g. the C57 strain).

Good breeding is as old as human society itself, but at four or five generations per century (or fewer), humans can barely keep up with changes in their environment! Mice, in contrast, have the advantage that they reproduce as quickly as three generations per year, allowing specific characteristics to be rapidly selected for. As there are believed to be fundamental patterns of behaviour that are conserved across all mammalian species, including mice and humans. A research program might ask whether 'equivalent' socially dysfunctional animal behaviours such as poor grooming, social withdrawal or dysphoria (all criteria for schizophrenia) may arise from the social environment, or purely as a result of the animal's genetic background.

Breeding an imperfect mouse

Mice may, therefore, be housed in either enriched (toys, terrain, mazes, and plants) or featureless environments (a plain cage) and assessed for their social ranking behaviour in both environments, and bred according to their specific behavioural characteristics. Mice that succeed and mice that do not succeed in each environment (as defined below) may be interbred, especially according to whether they display anti-social dominant (aggression) or sub-dominant behaviours (withdrawal) or sociable dominant/subdominant behaviour. These behaviours may be assessed as poor grooming, aggressive behaviour, a lack of environmental 'opportunism', withdrawal, persistent unwanted behaviours (e.g. mounting an unreceptive female) etc. After ten generations, the behavioural responses of the eleventh generation litters, raised either in isolation or communally, may be tested in response to their introduction to a novel environment or to new individuals.

Behaviours that may be selectively bred for include aggression, withdrawal, poor grooming, anxiety, dominance and environmental niche colonizing capacities (e.g. the exploitation of a new area within a 'cage' accessible only by swimming or climbing). As additional 'control', mice that have been acutely isolated, or kept in impoverished environment or within overcrowded conditions for one generation (from original strain) will be used to dissect genes that change in expression over one lifetime from changes that associated with 'genetic drift' due to selective breeding (within or between strains).

Their behaviours will be assessed in response to each of the breeding populations to see if the nature of the individuals from each selected breeding population affects or predicts their specific behaviours. Where there are reproducible differences in behaviours between the breeding extremes (socially dominant, socially recessive or poorly groomed etc.) the nervous tissue from the eleventh generation *post-mortem* will be exposed to two dimensional gel electrophoresis to ascertain whether there are any quantitative differences in the amounts of proteins (levels of gene expression), properties (gene processing) or sequence (mutations or polymorphisms) present in the brain that may be associated with changes in brain function and behaviour. A negative finding may support the contention that there is no genetic basis for social dysfunction or rejection, instead of supporting the argument that such behaviours are 'learned' or conditioned by the animal's social environment.

This research will thus address the eugenic basis ('good genes') for social dominance hierarchies and breeding patterns when compared to unselected breeding populations which have been allowed to breed undisturbed within an enriched or an impoverished (featureless) environment. The associated behavioural studies described above will seek to establish patterns of eleventh generation mouse behaviour under the four environmental conditions chosen (enriched and crowded vs. enriched and sparsely-populated vs. impoverished and crowded vs. impoverished and sparsely-populated), for mice that have been selectively bred for extreme social characteristics and for 'control' mice that have been taken from the 'normal' (wild-type) breeding population.

It is hoped that these studies will ascertain whether aspects of environment, for example housing density (stress from overcrowding) or an impoverished environment, may trigger antisocial or aggressive behaviours in mice from each of the selectively bred stocks, and to determine which of these behaviours, if any, may be inherited, if so to what extent in each of the selectively bred populations.

Proteomic analysis of the inheritance of behavioural traits

After breeding and exposing both control animals and specially bred animals to certain environments, the patterns of expression of the mouse genome in relation to its environment, the cell's 'protein complement', will be determined via new proteomic technologies screening, *as the protein is the final functional manifestation of a gene in response to a cell's interaction with its global environment.*

Firstly, the various regions of the brain are separated, then the regions of the cells therein are separated into nuclear, soluble, membrane and other subcellular fractions. The proteins so isolated may be run and separated in two further dimensions, firstly by an electrical field across a gradient of pH to separate the proteins by virtue of their inherent net electrical charge. The proteins may then be further separated in a fourth dimension by denaturing them in a detergent which confers an equal distribution of negative charge (SDS + DTT) and then allows them to separate according to physical size, the largest moving the most slowly within the imposed electrical field.

This results in a gel showing the relative abundance, size and charge of each protein by region and cellular compartment. By comparing gels it is possible to see what has changed between animals that have been differentially bred and treated. The proteins that have changed in their properties or abundance can be extracted and sequenced to determine their differences in genetic sequence and post-translational modifications. Hence we may be able to infer whether an individual's response to and success within an environment, isolated or densely-populated, engaging or Spartan, is 'indelibly inscribed' within his or her genes, or whether even an animal from 'inferior' stock may, by virtue of its interaction with its environment, be able to 'succeed' irrespective of inherited traits that appear predict otherwise.

APPENDIX K

Seeing a change against the light: how neural circuits are adapted in the retina

Philosophers have long argued whether we perceive or truly observe our surroundings, but in a world fraught with challenge and opportunity the selective advantage of a visual system is evident throughout nature. The evolution of the retina has allowed us to detect the light from a star, contrast images in the desert sun and discern subtle movements in a bustling city street. It has taken hundreds of millions of years to refine the structure and function of the human retina. Yet the original blueprint was so successful that the fundamental architecture of the retina has been conserved in all species from fish to primates. Not only does the retina relay information about light, contrast, and movement, but all the more remarkably it is able to do so whether in the twilight or sunlight through a process known as adaptation. Yet only recently have the intricate cellular processes that underlie adaptation been unravelled, revealing how the information neurons carry is altered in response to an ever-changing environment. Such developments take us a few steps closer to one of the goals of neuro-ophthalmologists - the design of an artificial retina. But the retina is more to science than a study of vision, it has become a window to understanding how cells communicate with one another and the world outside.

Light from our surroundings is collected by the eye through a regulated aperture called the iris and focused as an inverted image onto the retina. It is this simple physical arrangement that is captured in the design of the camera, and yet science is far from designing a light detection system as sensitive and as detailed in its resolution as the retina. Contrary to popular conception, the retina is, in fact, an integral part of the brain and has provided a focus for the study of the nervous system since Ramon Cajal first described the architecture of the tissue around the turn of the century. Since his pioneering work science has tried to relate the structure of the retina to its function, and more recently to understand the biochemical and biophysical basis of light adaptation. As an electrical circuit, the retina is perhaps one of the best understood parts of the nervous system. The function of the retina, as for any given circuit, is defined by the precise arrangement of chemical switches and electrical contacts that are formed between individual nerve cells, or neurons. The resulting neuronal networks serve to direct sensory information to centers that first integrate and then pass on the processed information to the appropriate command centers that control the muscles, glands and organs. Following this general plan the retina consists of a layer of specialised sensory neurons called photoreceptors that capture the light image and convert it into electrical information which is then transmitted 'vertically' to a layer of bipolar relay neurons. It is at the level of the bipolar cells that two parallel pathways of information flow are created. One population of bipolar cells is excited when the intensity of illumination is increased and the other responds when it is decreased, and the two populations are referred to as On and Off bipolar cells respectively. The On and Off bipolar cell populations channel information about changes in light intensity onto a third layer of neurons known as ganglion cells. The ganglion cell layer serves as a centre for the integration of light information in the retina, where subpopulations of On and Off ganglion cells respond with a burst of spikes in response to increases or decreases in the intensity of light respectively, depending upon the nature of the bipolar cell from which they receive their input.

So why did evolution take on the apparent expense of creating distinct On and Off channels for the processing of light information in the retina? This arrangement confers two advantages, in the generation of a signal as an object moves across the field of view and in the enhancement of contrast between areas of relative darkness and light. Ganglion cells integrate all the information they receive from On and Off channels and encode their computed output in the form of electrical spikes which travel along the optic nerve to another relay centre in the midbrain, the lateral geniculate nucleus of the thalamus. From this point, light information is directed to the visual cortex and other centres in the brain for further processing and image reconstruction, completing the description of the 'through' pathway for information flow. One of the reasons for our high level of visual acuity is our ability to contrast between areas of relative darkness and light. This capacity is achieved in part by the lateral processing of light information within the retina. The light signal from the photoreceptor is not delivered exclusively to the bipolar cell layer, as information also flows to a layer of horizontal cells which lie between the bipolar and photoreceptor cell layers. Horizontal cells are believed to spread a radiating wave of inhibition to surrounding bipolar cells by releasing the inhibitory transmitter gamma-aminobutyric acid, or GABA upon stimulation. The signal sent by the horizontal cell layer to the bipolar cell is in effect the reverse of the information about the amount of available light that it receives from the photoreceptor. So where the photoreceptor senses light, the horizontal cell network informs the surrounding On and Off bipolar cell pathways that they are relatively dark and vice-versa, providing us with an exaggerated impression of contrast between regions of light and shade in our surroundings. In more philosophical terms, where we see black and white there may in fact only be shades of gray. The artist Claude Monet took the idea of contrast one step further, and was renowned for his striking use of contrast enhancement between objects in his paintings. Through an absence of natural shading in his paintings, he was able to create a dramatic impression of contrast, in effect imitating the function of our retina.

Further lateral processing occurs in the inner retina, where a heterogeneous population of amacrine cells receive information directly from the bipolar cell layer and radiate a range of both inhibitory and excitatory outputs onto the receptive terminals of surrounding ganglion cells and the output terminals of the bipolar cell layer. In addition to the vertical and lateral processing of light information, a population of amacrine cells called interplexiform cells feeds information about the amount of light flowing in the retinal circuit back onto the receptive terminals, or dendrites of horizontal and bipolar cells in the outer retina by releasing the hormone dopamine. Dopamine diffuses widely and changes the sensitivity of retinal neurons to light to coordinate the process of adaptation. Another important role that has been proposed for amacrine cells is in the detection of movement. A population of amacrine cells in the salamander receives its information from both On and Off channels, and fires a volley of spikes in response to the beginning and end of a light stimulus, such as a target moving across the field of view, and it believed that an equivalent population of transient amacrine cells may exist in the mammalian retina.

It is only very recently that researchers have been able to describe the intricate cellular pathways that are involved in the detection and amplification of light signals and to elucidate the biochemical pathways that modify the size and shape of the responses of individual neurons to light. In the first step in phototransduction, the process of converting light energy into electrical energy, individual light particles, or photons are captured by pigments in the outer segments of photoreceptors. The rod photoreceptor has been the more extensively studied of the two and is specialised for nocturnal, or scotopic vision. When we refer to cones we tend to think of their familiar role in colour vision. Populations of red, green and blue cone photoreceptors are present at high density in the specialised foveal region of the retina, each containing a specific pigment that absorbs light over a narrow band of the spectrum of visible light. However, the principal role of the cone photoreceptor is in the measurement of light intensity rather than wavelength, and to this end the molecular design of cones enables them to operate at higher background light intensities, whereas rod photoreceptors are specialised for the detection of very low light levels. Although cones do not operate efficiently in the twilight and require the assistance of rod photoreceptors, their design allows them to distinguish changes in light intensity over a large range of background light levels to confer daylight, or photopic vision.

Light information from cones passes directly into the On and Off bipolar cell channels, whereas information from rod photoreceptors in the mammalian retina is sent exclusively to a population of 'rod' bipolar cells, which in other respects resembles the cone On bipolar cell in the nature of its responses to light. This would at first glance suggest that the rod pathway only serves to detect the presence of light rather than generating On and Off channels to gain the advantages of contrast and movement detection. Ultimately however, the rod pathway does separate the night-time picture into distinct On and Off channels that the visual system can interpret, and does this by feeding the output into the On and Off channels created by the cone bipolar cells, thus avoiding the loss of resolution and spatial complications that the creation of two additional rod channels would incur. The rod bipolar cell funnels all of its information about increases in light intensity into the AII amacrine cell, a specialised neuron that then feeds this information directly into the On and Off channels. In response to light the AII amacrine cell simultaneously decreases information flow through the Off-channel, whilst increasing information flow in the On channel, and this feat of contradiction is made possible because the AII amacrine cell employs both chemical and electrical methods of information transfer within a single type of neuron. Upon illumination, the

AII cell releases the inhibitory transmitter glycine onto the output terminals of the cone Off bipolar cell and passes excitatory information directly into the cone On bipolar cell because the two cells are electrically coupled as if the two were wires connected to a closed switch. This electrical switch is created by the presence of conductive pores known as gap junctions adjoining pairs of cells which allow charged ions to pass freely between cells, and so excitation in one neuron is transmitted directly into another without the need for the release of a chemical intermediary. Gap junctions also connect pairs of horizontal and AII amacrine cells and assist in the lateral processing of light information through these electrically coupled networks. In a far-reaching recent discovery Stephen Mills and Stephen Massey showed that the hormone dopamine, released by certain amacrine cells in response to light, closes down the gap junctions between pairs of AII amacrine cells as well as those between pairs of horizontal cells, and dopamine does this by increasing levels of the intracellular messenger cyclic adenosine monophosphate (cyclic AMP) in these neurons. Mills and Massey were further able to demonstrate that another messenger molecule released from cells in the retina in response to increases in background light, the infamous nitric oxide, specifically closes gap junctions that electrically couple the AII amacrine cell and the cone On bipolar cell. The closure of this gap junction has the interesting consequence of shutting down the pathway for night vision. These exciting findings show that the retina literally 'rewires' itself from nocturnal to diurnal vision through the release of specific hormones in response to an increase in the ambient level light, as part of the integral process of adaptation.

Many people are familiar with the idea that most neurons encode electrical information in the form of spikes or action potentials. Bipolar cells release the excitatory transmitter glutamate from their presynaptic terminals onto the postsynaptic dendrites of amacrine and ganglion cells in response to changes in light intensity. Glutamate binds to a receptor on the postsynaptic membrane where it opens pores which allow positively charged sodium ions to flood into the dendrites, making the ganglion or amacrine cell more conductive in terms of information flow. These pores are better known as ion channels and are found in all types of cell. Ion channels are frequently selective for certain ions, such as chloride, sodium or potassium, and control the distribution of these charged ions across the cell membrane determining the voltage, or potential across the cell membrane. As the membrane potential determines the state of activity in cells, especially neurons, they play a central role in cell signalling in almost all processes. The influx of sodium ions triggered by the binding of glutamate to its receptor results in a run-away spike depolarization, or action potential because a population of voltage-sensitive sodium channels are rapidly activated by the depolarization once the membrane potential has crossed a certain threshold. The depolarizing phase of the action potential is then quickly reversed by a reflexive outward movement of potassium ions returning the membrane potential a less excited state. This wave of membrane depolarization propagates along the length of the output process of the neuron to the presynaptic terminals of the neuron, where it opens excitable calcium channels. The resulting influx of calcium ions into the terminal causes the release of transmitter from specialised membrane storage compartments, which are triggered to fuse with the pre-synaptic membrane and release their contents into the synaptic cleft by the entry and binding of calcium to special sensor proteins in the terminal.

The opening of excitable sodium and potassium channels to generate the depolarizing waveform known as an action potential was first described by the Nobel Laureates Alan Hodgkin and Andrew Huxley, but since their pioneering efforts we are now aware that information is not only encoded in the frequency of these action potentials, but may also be encoded in the delay and the shape of the action potential. Whilst ganglion cells and most amacrine cells encode and pass on light information received from bipolar cell pathways through the generation of action potentials, the cells of the outer retina are unusual in that they possess no excitable population of sodium channels that are opened by membrane depolarization. Photoreceptors, bipolar cells, and horizontal cells respond to light with only slow, graded changes in their membrane potential and such graded changes in membrane potential migrate more slowly and less efficiently to the terminal. It took some time before Craig Jahr, George Ayoub, and David Copenhagen were able to show that information flow from photoreceptor terminals is also mediated by the release of glutamate onto the receptive dendrites of bipolar and horizontal cells. Illumination of the photoreceptor causes a graded reduction in the rate of release of glutamate from the photoreceptor terminal. David Attwell and colleagues subsequently revealed that it is this reduction in glutamate release that causes the Off bipolar cell to become less conductive of light information and On bipolar cells to become more conductive. The observations by Attwell, Wilson, Scott Nawy and David Copenhagen that the nature of the responses to light and glutamate in the On bipolar cell is the opposite of those of the Off bipolar cell suggested that glutamate activates very different signal transduction pathways in the two types of bipolar cell.

Action potentials stimulate transmitter release from presynaptic terminals in an explosive 'all-or-none' fashion, and little release of transmitter tends to occur before the arrival of an action potential, whereas the graded potentials of photoreceptors and bipolar cells cause a graded change in the rate of continuous release of glutamate from the terminals of these neurons that is in some way proportional to the size of the light stimulus. The encoding of light information in these neurons is therefore entirely relative. In the dark-adapted retina, a few photons of starlight can stimulate a small yet perceptible change in the release of glutamate from the terminals of photoreceptors and bipolar cells, whereas a comparatively larger increase in light intensity may not elicit a perceptible response in daylight. This is the underlying principle of light adaptation, where the retina switches from a detector and amplifier of light to a circuit that is comparative, measuring relative changes in light intensity. Cells in the outer retina are therefore not so much designed to pass on information about the absolute intensity of light at a point in the field of view, but to be able to make measurements of relative changes in light intensity throughout a range of background levels of light.

We can perhaps now begin to appreciate how the process of phototransduction serves to encode changes in the intensity of light into electrical signals that represent the language of the nervous system. The outer segments of rod photoreceptors contain the light-capturing pigment rhodopsin, which is composed of the chromophore 11-*cis*-retinal chemically bound to the protein opsin. Rhodopsin is present in the membrane discs of the outer segment, spanning the membrane seven times in all. Selig Hecht showed as long ago as 1938 that the absorption of a single photon of light by a human rod cell is sufficient to cause a perceptible one milliVolt change in the potential across the photoreceptor membrane. This was later shown through the classical experiments of George Wald and co-workers to result from a photoisomerization, or geometrical change in the structure of 11-*cis*-retinal to *trans*-retinal, converting the energy of light into atomic motion. This atomic motion is in turn transmitted through the chemical bond linking retinal to opsin, causing a structural change in its protein partner within milliseconds. The activated form of the protein, metarhodopsin is then in the right 'shape' or conformation to activate a universal cellular signalling mechanism which translates the atomic motion of receptors into changes in the energy stored in chemical bonds. This mechanism is provided by 'G'-proteins, which are in essence a type of 'chemical switch' that couples signals received by receptors in the cell membrane into biochemical processes that take place both inside the cell and elsewhere in the cell membrane. G-proteins are found anchored to the membrane when unstimulated and are formed from two components; an alpha subunit and a seemingly inseparable combination of beta and gamma subunits. Metarhodopsin causes the alpha subunit to exchange the nucleotide guanosine diphosphate (GDP) for guanosine triphosphate (GTP) which contains an extra high energy bond between its second and third phosphate molecules. The presence of this additional negatively charged phosphate bound to the alpha subunit commands a change in the conformation of the alpha subunit which causes it to lose its affinity for the beta-gamma subunit. The two components then separate to form a bifurcating, or forked signalling pathway. The alpha subunit, now in its stimulated GTP-bound state is then able to activate the enzyme phosphodiesterase by pulling away its inhibitory subunit.

The phosphodiesterase of the photoreceptor cleaves the high energy phosphate bond that is formed within the second messenger molecule cyclic guanosine monophosphate, or cyclic GMP which is produced by the action of the enzyme guanylate cyclase upon GTP. A second messenger is any ion, molecule or protein that mediates the transfer of information within a cell following the arrival of a signal carried by a primary messenger to the cell surface, which may be a hormone, transmitter or in the case of the photoreceptor, light. The regulation of cGMP levels in the photoreceptor reflects the essence of many second messenger pathways where a metabolic building block such as the nucleotide component of GTP used in the assembly of DNA, is transformed through a one-step dead-end biochemical pathway into a short-lived messenger molecule that is itself a highly specific and carefully regulated signal. This is elegantly demonstrated by the process of converting GTP into cyclic GMP through the action of a guanylyl cyclase and then back into the 'linear' non-messenger form guanosine monophosphate (GMP) by cleaving the phosphate bridge formed by the phosphate molecule by the action of a phosphodiesterase enzyme, the entire process being driven by the release of free energy stored in the bonds formed between phosphate molecules. The concentration of cGMP becomes the critical signal informing the visual system as to relative changes in light intensity in the photoreceptor at a point in time, and is itself precisely regulated by the balance of phosphodiesterase and guanylate cyclase enzyme activities in the photoreceptor. However, if changes in the cGMP concentration provide an almost instantaneous measure of changes in relative light intensity, other signalling pathways must have evolved to ensure that the sensitivity of elements that detect and control the concentration of cGMP are continuously adapted, or modulated to the absolute background level of light.

But what role do the beta-gamma subunits play in the light response? It was once thought that they served only as anchors for alpha subunits in the membrane, but they have now been implicated as major players in a number of signalling pathways. In the photoreceptor dissociated beta-gamma subunits migrate to activate rhodopsin kinase which is instrumental in the termination of the light response. Rhodopsin kinase belongs to another class of enzymes involved in cell signalling called protein kinases which participate in another ubiquitous 'dead end' signalling pathway of cell metabolism. Protein kinases facilitate the transfer of the terminal phosphate molecule from adenosine triphosphate (ATP), the universal energy currency of the cell, onto specific amino acids of the target protein. These are amino acids which possess 'water-like' hydroxyl (-OH) groups and are surrounded by a specific signal sequence of amino acids that are recognised by the kinase before the transfer of a phosphate molecule onto the amino acid can occur, a process known as phosphorylation. Protein kinases are classified into two broad families according to whether they phosphorylate the hydroxyl groups of the amino acids serine or threonine, like rhodopsin kinase, or the larger amino acid tyrosine, an event typical of many pathways involved in the regulation of cell growth and division. The transfer of a highly negatively charged phosphate molecule onto a target protein has several effects. Firstly it dramatically alters interactions between charged residues within the target protein resulting in a change in its conformation and consequently its activity, and secondly it alters its interactions with other proteins either as a result of these conformational changes or from direct interactions with the added phosphate groups. In some proteins, phosphorylation alters the affinity of a protein for the cell membrane, which may sometimes lead to a change in its location within the cell.

This does not fully explain how rhodopsin kinase inactivates metarhodopsin. Once rhodopsin kinase has performed multiple phosphorylations at the terminal portion of metarhodopsin, another protein aptly named arrestin binds to metarhodopsin to prevent it from further stimulating G-protein activity. This, in cell terminology, is known as a feedback pathway, where the duration of a signal is regulated by downstream elements in the pathway; in this case, the receptor that stimulates the G-protein is itself switched off by an element downstream of G-protein activation. The light response finally ends for metarhodopsin when it is regenerated by the replacement of the spent *trans*-retinal chromophore with fresh 11-*cis*-retinal, whereupon arrestin can no longer bind to phosphorylated rhodopsin and leaves. A protein phosphatase, the reverse component of the 'dead-end' kinase signalling pathway, is then able to strip the phosphates off the terminal portion of rhodopsin in preparation for another cycle. As for the alpha subunit, it possesses an intrinsic time switch which is triggered by its interaction with the cGMP phosphodiesterase, resulting in the cleavage of the terminal phosphate of its bound GTP molecule in a second example of a feedback pathway. This returns the alpha subunit to its inactive GDP-bound conformation and allows it to reassociate with beta-gamma subunits, and the G-protein cycle is ready to start again.

But how is the concentration of cyclic GMP finally translated into an electrical signal? It seemed reasonable to predict that the photoreceptor possesses a sensor that couples changes in the cyclic GMP concentration to the observed changes in membrane potential. Fesenko and colleagues were the first to be able to show that cGMP binds directly to a population of ion channels causing them to open. These cGMP-regulated channels were shown to be selectively permeable to calcium and sodium ions, which in the absence of light-stimulated phosphodiesterase activity flood into the cell and depolarize the photoreceptor membrane. This depolarization is, in turn, is propagated and translated into an increase in the rate of glutamate release from the photoreceptor terminal. Conversely, in response to light phosphodiesterase reduces the cyclic GMP concentration, cyclic GMP-regulated channels close, the intracellular face of the membrane becomes more negatively charged, or hyperpolarized and there is a subsequent decrease in glutamate release from the photoreceptor terminal. So how does a single photon generate as much as a milliVolt change in membrane potential? The answer lies in the remarkable amplification of the light signal by the elements of the phototransduction cascade. A metarhodopsin receptor activated by a single photon has been estimated to activate about five hundred G-proteins during its lifetime. In turn, each phosphodiesterase enzyme activated by a G-protein can cleave more than four thousand molecules of cGMP for each second of its active lifetime. The resulting fall in the cGMP concentration due to the activation of this biochemical cascade is sufficient to close hundreds of cation channels and evoke a membrane hyperpolarization as large as a milliVolt, which we sense as a reduction in the rate of release of glutamate from the photoreceptor terminal.

But what happens to the calcium ions that flood into the photoreceptor outer segment through the cyclic GMP-regulated channel? Calcium in its free ionic form is normally kept at very low concentrations inside cells, in fact as much as one hundred thousand times lower than the external concentration. In the photoreceptor, this gradient is maintained by a membrane transport protein that continuously exports calcium and potassium ions from the cell interior in exchange for entry sodium ions from outside the cell. Following the influx of calcium ions through the cGMP-regulated channels specialised intracellular proteins bind to free calcium ions with high affinity to prevent the internal calcium concentration from reaching levels that will destroy the cell. King-Wai Yau and Nakatani were the first to show that this membrane transport protein continues to export calcium ions after the cGMP-regulated channels that mediate calcium entry have closed in response to light, which Peter McNaughton and colleagues subsequently showed causes a fall in the intracellular calcium concentration which is a critical signal in adaptation.

But at what points in the phototransduction cascade is the degree of amplification, or sensitivity to light controlled to provide a mechanism for adaptation to lower background levels of light? The first clue was provided by the observation that after a prolonged period in the dark calcium levels in the unstimulated rod photoreceptor become elevated due to the continuous influx of calcium through GMP-regulated channels. The team of Mark Gray-Keller, Arthur Polans, Kris Palczewski and Peter Detwiler demonstrated that the calcium-binding protein recoverin prolongs the duration of the light response by preventing rhodopsin kinase from phosphorylating metarhodopsin. Denis Baylor and Leon Lagnado showed that in addition to recoverin, a number of other calcium-regulated factors combine to amplify light-stimulated increases in phosphodiesterase activity as much as thirty-fold. However Koch and Stryer, and later Yiannis Koutalos and colleagues provided evidence that increases in free calcium that occur in dark-adapted rod photoreceptors also inhibit the production of cyclic GMP by guanylate cyclase through another unidentified calcium-binding protein, decreasing the size of the current carried through cyclic GMP-regulated channels. Lastly, but by no means finally, Yi-Te Hsu and Robert Molday showed that calcium binds to a ubiquitous calcium-regulated protein known as calmodulin, causing a reduction in the affinity of the channel for cyclic GMP. This, in turn, results in a decrease in the sensitivity of the channel to changes in cyclic GMP concentration and may serve to extend the range of cGMP concentrations over which the channel sensor operates. Combining the actions of these various calcium-binding proteins of the rod photoreceptor into the larger picture of adaptation, Yiannis Koutalos and King-Wai Yau were able to make a number of predictions about the consequences that a fall in intracellular free calcium levels will have upon the degree of amplification of the light signal in broad daylight. The decrease in free calcium levels will increase the cGMP concentration in the rod photoreceptor by both inhibiting phosphodiesterase activity and stimulating guanylate cyclase activity. More cation channels will open due to the elevated cGMP concentrations and their increased affinity for cGMP, because calmodulin is no longer activated by calcium. The combined effect of these decreases in calcium during light adaptation is to reduce the sensitivity of the rod photoreceptor to a given increase in light intensity, because each excited metarhodopsin protein hydrolyses a smaller fraction of the available pool of cyclic GMP molecules due to the reduced level of gain in the phototransduction cascade and the increased cGMP concentration. The net advantage that all of these biochemical adjustments confer is to trade a reduction in the absolute sensitivity of the rod photoreceptor to light for an increase the range of background light levels over which the photoreceptor is able to detect a change in light intensity.

Is calcium the only second messenger that plays a role in light adaptation in the photoreceptor? It would appear not. The guanylate cyclase present in the photoreceptor has a built-in sensor that detect increased nitric oxide levels, and when this sensor binds to nitric oxide it stimulates an increase in cGMP production. A team led by Steve Barnes and Dmitry Kurrenny provided evidence that nitric oxide opens both cGMP-regulated cation and excitable calcium channels in rod photoreceptors. Their findings strongly suggest that nitric oxide production in response to increased levels of background light serves to amplify changes in the rate of glutamate release from the rod photoreceptor by enhancing the rate of calcium entry into the terminal. Further, increases in cyclic GMP concentration stimulated by nitric oxide will help to extend the range of light intensities over which the photoreceptor is responsive to changes in light intensity.

Now that we have seen that adaptation to light involves both the closure of gap junctions and changes in the sensitivity of photoreceptors to light, are the bipolar cells that create the On and Off channels modulated during adaptation? To answer this question we need to understand the nature of the glutamate receptors present on the two types of bipolar cells that regulate the passage of information through the On and Off channels of the retina. Neurons have evolved two general types of receptor for the transmission of information at chemical synapses. The fastest of the two forms of receptor-mediated transmission at chemical synapses occurs when the transmitter substance opens an ion channel directly. As both the ion channel and the region that binds to transmitter are parts of the same assembly of protein subunits, conformational changes in the receptor are transmitted directly into the rapid opening of an ion channel. The second type of receptor transfers information from the transmitter into the stimulation of a G-protein pathway which then regulates one or more ion channels either directly or via an intermediate messenger such as cyclic GMP. Signals passing through such a metabotropic receptor pathway are therefore slower than those transmitted via receptor-operated ion channels, although metabotropic receptor pathways present more potential levels of regulation for the fine tuning of the synaptic response which may have been a key factor in their evolution. At the synapse between the Off bipolar cell and the photoreceptor, glutamate binds to a receptor-operated channel which is permeable to positively charged ions, or cations, and the resulting influx of these ions causes the postsynaptic membrane of the dendrites to become depolarized. The resulting wave of membrane depolarization travels to the bipolar cell terminal, resulting in the opening of calcium channels and an increase in the rate of glutamate release onto the receptive dendrites of amacrine and ganglion cells. Receptor-operated channels that selectively allow the entry of negatively-charged chloride ions into the cell are also present in most neurons in the retina. Such receptor-operated chloride channels bind to the inhibitory transmitters GABA or glycine that are released from horizontal and amacrine cells, and the resulting influx of negatively-charged chloride ions into neurons limits the extent of excitation in neurons bearing these receptors. In nature *ying* and *yang* are inseparable, and where excitatory receptor-operated channels are found, others that are inhibitory must be present to restore the balance, and in the brain, the absence of such a balance can result in an epilepsy.

The first important clue that adaptation to light might occur in the channel created by the Off bipolar cell was provided by Gilbertson and Wilson, who showed that the receptor-operated channel of the Off bipolar cell is of an unusual type. This channel is composed of an arrangement of protein subunits that allow not only sodium ions, but also calcium ions to enter into the dendrites. This is a property shared by glutamate receptor-operated channels of the hippocampal region of the brain which mediate long-lasting changes in the passage of synaptic information through the admission of calcium into the dendrites, a process that is believed to be central to the higher processes of learning and memory. As the concentration of intracellular calcium provides a hinge around which cell processes swing in almost all cells, it is tempting to propose a role for calcium in adaptation. Logic directs us to the glutamate-operated receptor-channel as the most likely molecular target for calcium-mediated changes in the amplification of light signals in the Off bipolar cell. Reinforcing this proposal Greg Maguire and Frank Werblin were recently able to show that dopamine increases the size of the cation current passing through the glutamate receptor-channel. In an elegant study they revealed that dopamine binds to its membrane receptor at the surface of the Off bipolar cell and stimulates a class of G-protein whose liberated alpha subunits migrate to activate the enzyme adenylate cyclase which converts ATP into cyclic adenosine monophosphate (cyclic AMP) in a second messenger pathway that is analogous to the guanylate cyclase-phosphodiesterase pathway. The second messenger cyclic AMP diffuses to the regulatory subunits of protein kinase A which has a high affinity for the messenger molecule. The change in conformation in the regulatory subunits that follows the binding of cyclic AMP causes the liberation of the catalytic subunits of protein kinase A which are then free to phosphorylate the glutamate receptor-operated channel. This phosphorylation increases the size of the cation current in response to a given discharge of glutamate from the photoreceptor. By acting through the cyclic AMP second messenger pathway dopamine amplifies the magnitude of receptor responses in the Off bipolar cell to light signals received from the photoreceptor. Thus dopamine effectively increases gain in the light signal through the Off-channel in daylight, which may compensate for the parallel decrease in the amplification of the light response in the photoreceptor.

This brings us at last to the mystery of the On bipolar cell, whose responses to light are a curious inversion of those of the photoreceptor and the Off bipolar cell. The conundrum of this 'sign-inverting synapse' was solved by Craig Jahr and Gertrude Falk who were able to show that the glutamate receptor of the On bipolar cell activates a G-protein cascade which appears at first glance to be identical to the pathway in the photoreceptor. Binding of glutamate to the mGluR6 receptor stimulates a G-protein cascade leading to the stimulation of another cGMP-cleaving phosphodiesterase and the closure of another population of cation channels that appear to opened by increases in the cyclic GMP concentration. The genetic make-up of this glutamate receptor was subsequently described by Nakajima and Nakanishi and the receptor was found to be present exclusively in the On bipolar cell of the retina. The glutamate receptor of the On bipolar cell was the sixth member of the family of metabotropic glutamate receptors which exert their effects through G-proteins to be cloned, and hence it was given the abbreviation mGluR6. Taking advantage of the unique presence of this receptor in the On bipolar cell, Masu and Nakanishi bred a strain of mice in which both copies of the mGluR6 receptor gene had been 'knocked out' or deleted at the level of the germ cell. In these mice, all responses to light in the On channel were abolished, but responses carried by the Off-channel were left unscathed, and despite significant visual impairments these mice were in fact far from blind. These intriguing experiments support the theory that the On and

Off channels were created for the generation of contrast and the detection of movement, as neither channel on its own appears essential for sight. The idea that the receptor pathways of the On bipolar cell and photoreceptor persisted were essentially similar persisted until Pedro de la Villa, Takashi Kurahashi and Akimichi Kaneko demonstrated that the properties of the cyclic GMP-regulated channel of the On bipolar cell were fundamentally different from those of the photoreceptor. A further contradiction arose from the experiments of Rhodri Walters and Scott Nawy which showed that although there was much evidence for a G-protein receptor kinase to terminate the response to glutamate, there was, in fact, no evidence for the binding of an arrestin protein.

But how is the On channel adapted to changes in background light levels? There seemed to be little food for thought until Masayuki Yamashita and Heinz Wassle reported that the channel current that is suppressed by glutamate in the On bipolar cell was very sensitive to the concentration of calcium ions outside the cell. These findings were further explained when Rhodri Walters and Scott Nawy showed that current flow through the cyclic GMP-regulated channel was diminished if phosphorylation by a protein kinase activated by the binding of calcium-calmodulin was prevented. Moreover, cGMP-regulated channels quickly close even in the presence of high concentrations of cyclic GMP if the phosphate groups on the channel are stripped off by a phosphatase. Therefore during light adaptation, the size of the cation current carried through the channel may increase because calcium ions enter the cell, most probably through the channel itself, and subsequently activate the calcium-calmodulin-regulated protein kinase which then phosphorylates and opens more cyclic GMP-dependent channels. Again, as in the case of the photoreceptor and the Off bipolar cell, calcium appears to play a pivotal role in mediating adaptation in the On bipolar cell. But what role does dopamine play in adaptation in the On channel of light information processing? Walters and Nawy recently showed that phosphorylation by protein kinase A prolongs the active lifetime of the receptor after the receptor binds to glutamate. This effectively increases the efficiency of coupling of the mGluR6 receptor to G-protein stimulation and activation of the cGMP phosphodiesterase, prolonging the closure of the cGMP-regulated channels and the duration of the light signal in the On pathway. In fact for the mGluR6 receptor to be able to close cGMP-dependent channels at all, it appears that it must first be phosphorylated by protein kinase A. This may not be the complete story of modulation of the mGluR6 receptor in adaptation, as Walters and Nawy were also able to show that lowering the concentration of free calcium ions inside the cell using the rapid calcium-capturing compound BAPTA, resulted in the uncoupling of the receptor from the phosphodiesterase, implicating a further role for calcium ions in receptor regulation.

Putting the On bipolar cell into the general picture of adaptation, we may predict that increases in free calcium and cyclic AMP levels in response to daylight will increase both the size of the current carried through the cyclic GMP-dependent channel and also the ability of the metabotropic glutamate receptor to suppress it, as the degree of coupling of the mGluR6 receptor to the cyclic GMP-phosphodiesterase is enhanced by both protein kinase A and elevations in the levels of cellular calcium. However, as increased levels of calcium appear to open more cyclic GMP-dependent channels during daylight, the potential number of channels that are available for closure in response to glutamate will also increase, extending the range of background light intensities over which the On-bipolar cell can measure changes in light intensity. But this does not explain why distinct populations of rod and cone bipolar cells evolved, one specialised for nocturnal and the other for diurnal vision. The rod On bipolar cell is very effective in the detection of photoreceptor signals generated in response to changes in light intensity at very low light levels, and Jonathon Ashmore and Gertrude Falk were able to show that light responses may be amplified as much as two-hundred fold at the synapse between the rod-bipolar cell and photoreceptor. In contrast, evolution may have directed the cone On bipolar cell to change roles from a high-gain amplifier at low background light levels to a variable amplifier that is regulated by second messengers such as calcium and cyclic AMP which allow the cone On bipolar cell to adjust to a range of background light levels.

In summary, the retina has neatly evolved into a self-regulating neural circuit that can be studied in isolation, which converts the energy of light into electrochemical signals that can be deciphered by higher centres in the visual system. The photoreceptor captures and amplifies the energy of light and passes on the information encoded as changes in the rate of glutamate release from its terminals onto a layer of bipolar relay neurons. Bipolar neurons both amplify the signal and separate it into parallel On and Off channels in response to increases and decreases in light intensity respectively. Bipolar cells then pass photoreceptor signals that have been shaded by the output of surrounding horizontal cells onto the ganglion cell layer which integrates outputs from both bipolar and amacrine cells. The ganglion cell layer converts information about incident light from the graded analogue signal it receives from the bipolar cell into a digital signal which is then transferred at speed to the visual cortex for reconstruction. The retina has employed three principal second messenger pathways to rewire its circuitry from the efficient detection of low-level changes in light intensity at night, to one that measures rapid changes in light intensity against the higher background light levels of the day. The release of dopamine from amacrine cells into the retina as background light levels increase, leads to the amplification of signals from the photoreceptor in both On and Off bipolar cells through the elevation of cyclic AMP levels. In addition, dopamine shuts down gap junctional communication between horizontal and amacrine cells which appears to be of importance in enhancing signal resolution at low light intensities. Nitric oxide may increase both the amplification of light responses in the photoreceptor and the range of light intensities over which the photoreceptor is able to operate. In addition, nitric oxide closes down the channel of night vision at the level of the AII amacrine cell. A third intracellular messenger calcium alters the sensitivity of the photoreceptor and bipolar cells to changes in light intensity to provide a rapid channel-mediated barometer of ambient light levels. Increases in intracellular calcium levels reduce the sensitivity of the photoreceptor to changes in light intensity and increase the level of gain at the synapse between the bipolar cell and the photoreceptor. In both types of cell increases in cellular calcium levels appear to extend the range of background light levels over which the circuitry can detect changes in light intensity in exchange for a reduction in the absolute sensitivity to light, or amplification of the light signal.

What other fundamental questions remain to be answered about mechanisms of light adaptation in the retina? After the adage of the goose and the gander, one might reasonably predict that cyclic GMP levels in the On bipolar cell will increase in response to nitric oxide, and conversely that photoreceptor responses to light may be somehow modulated by dopamine. And if the gap junctional communication between horizontal cells is regulated by dopamine, what role is there for calcium and nitric oxide in regulating adaptation in these cells? If the analogue output of the photoreceptor, horizontal cell and bipolar cell circuitry of the outer retina is dramatically altered during adaptation, how is the digitising output of the amacrine and ganglion cells modulated by these second messenger pathways? It is already known that responses to light and GABA of ganglion cells are modulated by dopamine. Further, nitric oxide-sensitive forms of guanylate cyclase are present in certain populations of amacrine and ganglion cells, suggesting that this second messenger pathway modulates the excitability of these neurons in response to an increase in background light levels. To date over twenty types of amacrine cell have been described, each with its own distinct pattern of wiring and cargo of neurotransmitter, and their respective roles in the processing of light information is poorly understood. Perhaps one of the greatest challenges facing neuroscience is the functional characterization of these highly specialised cells that are involved in the primary processing of light information.

In conclusion, we are learning many fundamental lessons about cell signalling and inter-cellular communication from the retina. We are beginning to gain an insight into how molecules, proteins and membranes come together to form electrical circuits that are more sensitive than any that man has so far devised. The retina is not only a window to determining the how higher centres of the nervous system may function, it is a bridge towards understanding the interaction between mind and environment. By discovering how sensory information is detected and encoded in the retina, reconstructed in the visual cortex and ultimately converted into changes at the level of the genes themselves, we hope to understand the processes of the nervous system that form the basis of memory and experience and define our very being.

APPENDIX L

Further reflections on the role and function of GABA receptors

GABA$_C$ receptors

The study of the homo-oligomeric ρ1 receptor channel confers several advantages in the elucidation of the fundamental mechanisms governing the kinetics of binding and gating of ligand-gated ion channels (Colquhoun, 1999). First, homo-oligomeric ρ1 receptors display little or no desensitization upon prolonged application of high agonist concentrations (Chang et al., 2000), effectively removing the complexity of desensitized channel states from the kinetic analysis of channel activation and deactivation. Second, the rates of activation and inactivation of ρ1 receptor channels following agonist addition or withdrawal are relatively slow in comparison to those of other ligand-gated ion channels, allowing a detailed kinetic analysis of binding and gating to be performed upon a single oocyte (Chang & Weiss, 1999).

GABA$_A$ receptors have a lower affinity for agonist than GABA$_C$ type receptors, but are considerably faster and exhibit desensitizing responses, thus mediating the rapid onset and termination of synaptic signals and are ubiquitous in their distribution throughout the nervous system (*De Blas, 1996*). The kinetically slower, higher affinity GABA$_C$ receptors have been shown to be present upon On-bipolar (*Feigenspan, Wassle & Bormann, 1993*), horizontal (*Qian & Dowling, 1993*) and cone photoreceptor cells (*Picaud et al., 1998*) of the retina, in addition to many classes of interneurons present elsewhere in the nervous system (*Schmidt et al., 2001*).

The agonist-binding domain of the GABA$_C$ channel is provided by subunits of the ρ receptor subfamily (*Lukasiewicz & Shields, 1998*). Following the cloning of the ρ1 subunit (*Cutting et al, 1991*), it was demonstrated to produce functional homo-oligomeric receptor-channels upon expression in Xenopus oocytes with a pharmacology related to that of GABA$_C$ receptors, though differing in other kinetic properties (*Wotring, Chang & Weiss, 1999*). Although hetero-oligomeric receptors formed by the co-assembly of white perch ρ$_{1B}$ and γ$_2$ subunits possess functional properties more reminiscent of wild-type perch GABA$_C$ receptors (*Qian & Ripps, 1999*), the stoichiometry of mammalian GABA$_C$ receptors remains undetermined. A detailed understanding of the kinetic properties of ρ1 receptor-channels at a molecular level (*Chang & Weiss, 1999*) is crucial not only in determining the molecular basis of the differences between GABA$_A$ and GABA$_C$ receptors in the temporal encoding of information, but also in evaluating the variations in the thermodynamics of agonist binding and channel opening at a molecular level.

In addition, the use of homo-oligomeric receptor-channels considerably simplifies the interpretation of the relationship between the primary structure of a ligand-gated receptor-channel and its biophysical and kinetic properties. Much attention has been focused upon the functions and governances of the various domains present within this ligand-gated ion channel receptor superfamily which includes the nicotinic, 5-HT$_3$, glycine, GABA$_A$ (α,β,γ,δ) and GABA$_c$ (ρ$_{1,2,3}$) receptors (Vafa & Schofield, 1998; Ortells & Lunt, 1995; Betz, 1990). The specific residues in the N-terminus responsible for agonist binding in GABA$_A$ (Amin & Weiss, 1993) and ρ1 homo-oligomeric receptor channels (Amin & Weiss, 1994) have been elucidated, as have specific residues within the second transmembrane domain (TM2) that contribute to the putative channel pore (Corringer et al., 2000), the selectivity filter (Corringer et al., 1999) and the transduction of agonist binding into channel gating (L301, Chang & Weiss, 1998). Further, residues within TM2 (Walters et al., 2000) and TM3 (Amin, 1999) which govern the sensitivity of GABA receptors to anesthetic action and barbiturates have also been characterized. To date however, no functional role has been elucidated for any residue within the proposed fourth transmembrane spanning domain.

Amin & Weiss (1996) empirically determined a requirement for three agonist molecules to bind to gate (open) the homo-oligomeric ρ1 receptor-channel, inferring a kinetic scheme with a minimum of three channel closed states. Moreover, a kinetic analysis of the slow and fast components of [^{3}H] GABA dissociation from ρ1 receptor-channels led to the suggestion that channel opening locks agonist onto the receptor, inferring that more than one channel open state might possibly exist (Chang & Weiss, 1999). By the substitution of a spectrum of residues at the putative interface between TM4 and the intracellular loop which mediates kinase regulation (Moss et al., 1992), a specific serine residue at 439 has been identified which governs the kinetics of receptor activation and deactivation by GABA in a concentration-dependent manner. The hydrophobicity and molecular volume of the side acid of the substituted residue at 439 are negatively correlated with the rate of receptor activation and deactivation, and slow channel deactivation sufficiently to reveal the existence of at least three open states, which are realized in a time and concentration-dependent manner upon exposure to agonist.

GABA$_A$ receptors

GABA$_A$ receptors are members of the ligand-gated ion channel superfamily which includes the nicotinic acetylcholine, 5-HT$_3$, GABA$_C$ and glycine receptors. This receptor superfamily share a common topology, with the N-terminus contributing to the ligand binding domain, 4 transmembrane spanning domains (TM1 to 4), an intracellular regulatory loop which receives phosphorylation signals and tethers the receptor, and a channel pore region which contains residues of from TM2. The considerable functional diversity of GABA$_A$ receptor-channels arises from a range of subunits which assemble to form functional ion channels – including α_{1-6}, β_{1-4}, γ_{1-4}, δ, ϵ, π and θ - by virtue of the alternative splicing and RNA editing of these subunits, and post-translational events including receptor clustering and phosphorylation.

In addition to GABA binding sites which are believed to be formed at the interface contributed by α and β subunits within the N-terminus, a number of key neuromodulatory binding sites are known to be present within the pentameric receptor-channel complex including high and low affinity binding sites for benzodiazepines (Walters et al., 2000), with the high affinity site formed within the N-terminus by residues contributed by both α & γ subunits. The modulation of GABA-mediated synaptic transmission thus underlies the pharmacotherapy of various neurological and psychiatric disorders including anxiety and epilepsy.

GABA – the bipolar transmitter

Until recently the acronym GABA had become synonymous with inhibitory neurotransmission. A fresh spate of reports, however, has made this generalization feel increasingly uncomfortable. Even a decade ago, inhibition in the CNS was a seemingly simple affair, with synaptic inputs from GABAergic and Glycinergic neurons providing a calming counterbalance to excitation through glutaminergic, cholinergic and serotoninergic influences. GABA receptor classification was a matter of A, B and C, all of which were thought inhibitory in their mechanism (Bormann, 2000). Excepting further consideration of the metabotropic $GABA_B$ receptor, which mediates inhibition through the activation of K+ channels, or through the suppression of Ca2+ channel activity in neuronal membranes (Bowery & Enna, 2000; Couve et al., 2000), GABAA and $GABA_C$ receptors are ligand-operated ion channels which are gates for the bidirectional passage of Cl- (and HCO3-) ions. As under most experimental recording conditions the intracellular Cl- concentration, or activity [Cl-]i, is below the predicted electrochemical equilibrium potential†, $GABA_A$ and $GABA_C$ receptors are assumed to exert an inhibitory influence by mediating the GABA-gated influx of negatively charged Cl- ions across the cell membrane.

Intracellular Cl- activity is, however, the great variable of cellular electrochemistry, and whilst large electrochemical gradients for Na+ and Ca2+ entry are ubiquitously maintained, and the equilibrium (reversal) potential for K+ is usually near or beyond (i.e. more negative than) the observed resting membrane potential of the cell*, commonly encountered Cl- reversal potentials fall well within the operational range of membrane potentials (-80 to +80mV) observed in most excitable cells. Furthermore, it is easier and more energy efficient for a cell to establish and maintain a [Cl-]i which is above or below the predicted Cl- electrochemical equilibrium, than it is to maintain the high Na+, K+ and Ca2+ gradients using energy-hungry Na+/K+ and Ca2+ ATPases. Thus Cl- provides cells with a flexible alternative in regulating their excitability and in the generation of functional diversity within the nervous system. Whilst it is well known that intracellular chloride is accumulated above electrochemical equilibrium within fluid and electrolyte secreting epithelia (Walters et al., 2002) by a Na+K+2Cl- cotransporter (O'Brien et al., 1993), many reports are now accumulating to suggest that [Cl-]i is also accumulated above equilibrium within a variety of cell types which functionally express $GABA_A$ receptors.

GABA excitation and cortical development

During early development [Cl-]i is elevated above equilibrium within immature neurons, causing $GABA_A$ receptor activation to be depolarizing (Ben-Ari, 2002). In fact GABA assumes a depolarizing role in the early development of many types of neurons (Leinekugel et al., 1999), including those of the lateral superior olive (Balakrishnan et al., 2003). Within immature neocortical neurons this is achieved by increasing early expression of the Cl- accumulating Na+K+2Cl- cotransporter (NKCC1), and by delaying the expression of the Cl- extruding K+-Cl- cotransporter (KCC2, Ben-Ari, 2002; Yamada et al., 2004). Indeed GABAergic synapses are often formed before glutamatergic contacts, allowing developing neurons to be become excitable, permitting growth and synapse formation, whilst avoiding the potentially neurotoxic effects of an imbalance between GABA-mediated inhibitory and glutamatergic excitatory influences (Ben-Ari, 2002). By so delaying the formation of inhibitory influences, cortical networks may later be fashioned in a competitive and activity-dependent manner, after the initial non-competitive establishment of the cortical web.

GABA – a variable neurotransmitter?

Whilst it has been known for some time that GABA may be depolarizing at early stages of development (Cherubini et al., 1991), many different types of fully differentiated adult cells have now been shown to be excited by GABA. Indeed a cluster of recent studies has firmly demonstrated that this phenomenon is not restricted to developing neurons, and may be plastic in nature. Many findings have demonstrated that GABA has a depolarizing action upon mature cortical neurons (for review see Stein & Nicoll, 2003). For instance $GABA_A$ may transiently act as an excitatory (depolarizing) transmitter after intense bursts of $GABA_A$ receptor activation within the adult brain. This appears to be due to an activity-dependent shift in the Cl- reversal potential, for example in mature hippocampal CA1 pyramidal cells receiving intense synaptic inputs (Isomura et al., 2003). Such GABAergic excitation participates in the expression of seizure-like rhythmic synchronization ("after discharge") in the mature hippocampal CA1 region (Fujiwara-Tsukamoto et al., 2003). It appears that the depolarizing potentials observed during epileptiform activity induced within hippocampal slices reflects glutamatergic activity and GABAAergic inputs from both stratum oriens-alveus interneurons (Perez Velazquez, 2003) and the stratum pyramidale (Sun & Alkon, 2001). This, of course, has profound implications for the treatment of epilepsy with GABAergic agonists (Ashton & Young, 2003).

GABA mediated excitability may however not be confined to pathophysiological epileptiform activity, and occurs in adult rat hippocampal CA1 pyramidal cells at theta frequencies (5-10 Hz) associated with GABAergic postsynaptic depolarization and a shift of the reversal potential from the equilibrium potential of Cl- towards that of HCO3-, whose electrochemical gradient is regulated by carbonic anhydrase activity (Sun et al., 2001). This theta activity was abolished by $GABA_A$ receptor antagonists and also by carbonic anhydrase inhibitors, but was largely unaffected by blocking glutamate receptors (Sun et al., 2001). In fact, spatial learning in a water maze appears to be regulated through such a reversal in the polarity of GABAergic postsynaptic responses by increases in carbonic anhydrase activity (Sun et al., 2001). Carbonic anhydrase also functions as a molecular switch in the development of synchronous gamma-frequency firing (40 Hz) of hippocampal CA1 pyramidal cells (Ruusuvuori et al., 2004). As a synaptic molecular switch, carbonic anhydrase appears to change the function of the GABAergic synapses from that an excitation filter to that of an amplifier (Sun & Alkon, 2001), a switch which appears critical for gating the synaptic plasticity that underlies spatial memory formation, and which may also enhance perception, processing, and the storing of temporally associated signals (Sun & Alkon, 2001).

It is hypothetically reasonable to argue, due to the fine and complex architecture of neurons, that asymmetric intracellular chloride activities may be created within the dendrites, axons and synaptic terminals of many neurons due to highly localized variations in Cl- fluxes through ion channels, and by the asymmetric activities and expression of Cl- transport mechanisms and carbonic anhydrase activity, even whilst GABA maintains a hyperpolarizing action at the soma of the same neuron. Indeed the activation of presynaptic GABAA receptors depolarizes presynaptic glutamatergic nerve terminals projecting to ventromedial hypothalamic (VMH) neurons, thereby facilitating spontaneous glutamate release by activating excitable Na^+ and Ca^{2+} channels (Jang et al., 2001). These regional elevations in $[Cl-]_i$ are generated by the activity of NKCCs and were responsible for the GABA-induced presynaptic depolarization (Jang et al., 2001). It is perhaps inevitable that many more such examples will be found. Some unpublished reports have suggested that longer axons may be depolarized by GABA (Robert Halliwell, University of Durham, UK).

Excitation of glia by GABA

Glial cells are known to interact extensively with neurons within the brain, and strongly influence their activity. Astrocytes, which are associated with synapses, serve to integrate neuronal inputs, and actively release transmitters which modulate synaptic sensitivity (Hansson & Ronnback, 2003). Glial cells not only participate in the formation and rebuilding of synapses, but also play a key role in neuroprotection and the restructuring of nervous tissue following injury (Hansson & Ronnback, 2003). Astrocytes and glial cells express many of the same ionotropic receptors as neurons, and NG2 cells engage in rapid signalling with GABAergic neurons through direct neuron-glia synapses (Fraser et al., 1994). Reports are beginning to emerge that the activation of $GABA_A$ receptors in such glial cells may also be excitatory. Indeed $GABA_A$ receptor activation depolarizes both proteoglycan NG2 expressing cells (Lin & Bergles, 2004) and acutely isolated hippocampal astrocytes (Fraser et al., 1995).

GABA evokes hormone release from neuroendocrine cells

GABA$_A$ receptors are expressed upon a wide range of neuroendocrine cells including insulin-secreting pancreatic B-cells (Glassmeier et al., 1998) and catecholamine releasing adrenal chromaffin cells (Peters et al., 1989; Walters et al., 2002a). In both these neuroendocrine cell types GABA is known to evoke a depolarization as [Cl-]i is maintained above equilibrium. This of particular interest in pancreatic Beta-cells which possess a pool of GABA-containing synaptic-like microvesicles, which are distinct from the population containing insulin granules (Sorenson et al., 1991). In human insulinoma cells, which also express functional GABA$_A$ receptors, GABA application evokes a membrane depolarization which activates excitable calcium channels and induces insulin secretion (Glassmeier et al., 1998). This GABA evoked depolarization is also observed in pancreatic beta-cells which have been engineered to express GABAA-receptors at supranormal densities (Braun et al., 2004). It, however, remains unclear whether the quantities of GABA normally released from this vesicular pool within the pancreatic Beta-cells would be sufficient to depolarize the Beta-cell via its autoreceptors to the threshold of voltage-gated Ca2+ channel activation which would evoke insulin secretion, or whether GABAergic afferents instead might trigger insulin secretion under physiological conditions. It has been shown that GAD-like immunoreactivity, indicative of the presence of GABAergic afferent innervation, is present only upon the insulin-secreting cells in the rat and mouse pancreas (Gilon et al., 1991). Irrespective of whether the mechanism is via autoreceptors or by direct synaptic input, or both, GABA would appear likely to evoke the release of insulin in vivo. Thus pharmacological modulators of GABA$_A$ receptor activity such as ethanol, neuroactive steroids or benzodiazepines may significantly alter patterns of insulin release.

Perhaps the most dramatic illustration that GABA serves as an excitatory transmitter in adult tissues, is the presence of functional GABA$_A$ receptors upon catecholaminergic adrenal chromaffin cells of the adrenal medulla (Peters et al., 1989). In the adrenal chromaffin cell the application of GABA elevates free [Ca2+]i (Doroshenko, 1989) by evoking membrane depolarization (Busik et al., 1996) to the known activation threshold for voltage-gated Ca2+ channels (Doroshenko, 1989). This indicates not only that [Cl-]i is accumulated above equilibrium in chromaffin cells, but also that an increase in GABAergic activity might be expected to augment or evoke catecholamine release. Immunohistochemical analysis has indicated the presence of both GABAergic afferent fibres and GABA-containing chromaffin cells in canine adrenal glands (Kataoka et al., 1986). A dense network of GABAergic fibers is present both at the boundary between medullary and cortical cells and within the medullary tissue itself. GABAergic fibres are observed to surround the chromaffin cells, some of these fibers entering the adrenal medulla together with splanchnic cholinergic nerves (Kataoka et al., 1986).

Adrenal medullary chromaffin cells, along with adrenaline (epinephrine)-producing cells of the medulla oblongata, are one of only two cell types in the body known to functionally express the full complement of enzymes of the catecholaminergic pathway. This pathway converts the amino acid L-Tyrosine sequentially via L-Dopa (Tyrosine Hydroxylase), Dopamine (Dopa DeCarboxylase), Noradrenaline (*norepinephrine*, Dopamine Beta-Hydroxylase) into adrenaline (epinephrine, phenylethanolamine N-methyl-transferase). Adrenaline is the primary catecholamine released from the adrenal medulla in response to stimulation and is responsible for increases in neural activity (alertness), cardiac output and blood pressure (e.g. Graham, 1990). Moreover, the expression of the enzymes of the catecholaminergic pathway is variably regulated by a number of hormones and transmitters, including insulin (Walters et al., 2002b). Thus the synthesis, the production and release of adrenaline is tightly controlled by many systems as it is the endocrine system's rapid response hormone.

The natural question would be to ask whether the stimulation of GABAergic afferents actually increases the release of adrenaline into the bloodstream. This, however, is a fait accompli, as the functional role of the GABAergic system in the regulation of catecholamine release from adrenal chromaffin cells has already been demonstrated. Kataoka and co-workers studied the catecholamine output of canine adrenal glands in situ using an autoperfusion system (Kataoka et al., 1986). They demonstrated that GABA modulates the spontaneous release of catecholamines, and that adrenaline release is elicited in response to electrical stimulation of the splanchnic nerve. Administration of $GABA_A$ receptor agonists such as THIP or muscimol increased the catecholamine content in adrenal effluent blood, whilst bicuculline, a GABAA receptor antagonist, reduced it. As a control baclofen, a GABAB receptor agonist failed to alter the catecholamine content of adrenal effluent blood, and denervation of the adrenal glands did not prevent the THIP-elicited release of catecholamines (Kataoka et al., 1986). Whilst these findings may have been received cautiously at the time, the plausibility of the contention that GABAA receptor activation is directly implicated in the stimulation of adrenaline release has gained weight. As GABA, along with acetylcholine, is effectively established as a major stimulator of catecholaminergic release, all of its classical modulators such as BDZs (Walters et al., 2000), ethanol and neuroactive steroids may also be expected to modulate adrenaline output, especially as a BDZ-sensitive alpha subunit has been shown to be expressed both in the adrenal and the PC12 cell line model (Walters et al., 2002a). This would at first appear to give rise to a potential therapeutic paradox, given that BDZs are employed clinically as anxiolytics. Given that adrenaline is the 'fast' stress hormone, then those therapeutic agents which target the GABA molecule might therefore also be predicted to influence stress physiology.

Clearly, GABA can no longer be simplistically viewed as an inhibitory neurotransmitter, for GABA does merely as the intracellular chloride activity directs. As far as evidence for a widespread excitatory role for GABA in the nervous and neuroendocrine systems is concerned, we have only recently started to look and thus we have only just begun to find. Perhaps most intriguing is the observation that the regulation of intracellular chloride activity provides cells with an efficient, plastic and very dynamic means of regulating changes in their excitability.

APPENDIX M

Bibliography

R.J. Walters. Ion channel regulation in small intestinal crypts. Ph.D. Thesis. (1993). Cambridge University Library.

Walters, R.J., Sepulveda, F.V. A basolateral K^+ conductance modulated by carbachol dominates the membrane potential of small intestinal crypts. Pflugers Arch. 1991 Nov;419(5):537-9.

O'Brien, J.A., **Walters, R.J.**, Sepulveda, F.V. Regulatory volume decrease in small intestinal crypts is inhibited by K^+ and Cl^- channel blockers. Biochim Biophys Acta. 1991 Dec 9;1070(2):501-4.

Walters, R.J., O'Brien, J.A., Valverde, M.A., Sepulveda, F.V. Membrane conductance and cell volume changes evoked by vasoactive intestinal polypeptide and carbachol in small intestinal crypts. Pflugers Arch. 1992 Sep;421(6):598-605.

O'Brien, J.A., **Walters, R.J.**, Valverde, M.A., Sepulveda, F.V. Regulatory volume increase after hypertonicity- or vasoactive-intestinal-peptide-induced cell-volume decrease in small-intestinal crypts is dependent on $Na(^+)-K(^+)-2Cl^-$ cotransport. Pflugers Arch. 1993 Apr;423(1-2):67-73.

R.J. Walters, F.V. Sepulveda. Ca^{2+} and voltage-dependence of K^+ channels in the crypt basolateral membrane. Biophys .J., *Abstract, W115, Baltimore 1996.*

F.V. Sepulveda, **R.J. Walters.** The stage of enterocyte differentiation determines resting potential and agonist responses. *Exp. Biol. '96, Washington D.C. abstract 3104,* FASEB J.

Stephens, L., Cooke, F.T., **Walters, R.**, Jackson, T., Volinia, S., Gout, I., Waterfield, M.D., Hawkins, P.T. Characterization of a phosphatidylinositol-specific phosphoinositide 3-kinase from mammalian cells. Curr Biol. 1994 Mar 1;4 (3):203-14.

P.T. Hawkins, A. Eguinoa, R-G. Qiu, D. Stokoe, F.T. Cooke, **R. Walters**, S. Wennstrom, L. Claesson-Welsh, T. Evans, M. Symons, L. Stephens. PDGF stimulates an increase in GTP-Rac via activation of phosphoinositide-3-kinase. *Current Biology*, **5**, 393-403 (1995).

Walters, R.J., Hawkins, P., Cooke, F.T., Eguinoa, A., Stephens, L.R. Insulin and ATP stimulate actin polymerization in U937 cells by a wortmannin-sensitive mechanism. FEBS Lett. 1996 Aug 19;392(1):66-70.

R.J. Walters, S. Nawy. Modulation of the retinal On-bipolar cell mGluR6 cascade by phosphorylation. *Soc. for Neurosci. abstract, 435.2, San Diego* 1995.

R.J. Walters. Seeing a change against the light: how neural circuits are adapted in the retina. *New York Science Times Article, June 1996.*

M.L.Chandler, I.H.Pang, R.Doshi, E.M.Wexler, **R.J.Walters**, S.Nawy, L.DeSantis, M.A.Kapin. *In vitro* and *in vivo* protective effects of eliprodil in the retina. *Investigative Ophthalmology and Visual Science* 1997, **38**, No.4, Pt1, p.725, *Abstract*

Walters, R.J., Kramer, R.H., Nawy, S. Regulation of cGMP-dependent current in On bipolar cells by calcium/calmodulin-dependent kinase. Visual Neurosci. 1998 Mar-Apr;15(2):257-61.

Nawy, S., **R.J. Walters**. Modulation by PKA of the responses to L-AP4 in retinal bipolar cells. *Soc. for Neurosci. abstract, 498.12, New Orleans,1997.*

Amin, J., Pollock, V., **Walters, R.J.** A novel ρ1 GABA receptor channel with slowed kinetics. *ARVO Abstract, Ft. Lauderdale, FL, May, 2000.*

Walters, R.J., Hadley, S.H., Morris, K.D., Amin, J. Benzodiazepines act on GABAA receptors via two distinct and separable mechanisms. Nature Neurosci. 2000 Dec;3 (12):1274-81.

Walters, R.J., Amin, J. Slowed kinetics of ρ1 receptors upon substitution at S439 correlate positively with both side chain volume and hydrophobicity (in preparation).

Walters, R.J. Synergistic activation of G- and L-type Ca2+ channels by D-glucose and membrane depolarization in pancreatic beta-cells. Soc. for Neurosci. Abstract (platform), New Orleans, 2000.

Brandt, M.V., **Walters, R.J.**, Willenberg, H.S, Scherbaum, W.A., Bornstein, S.R. Nerve Growth Factor Increases Expression of Tyrosine Hydroxylase in Serum-Starved PC12 Pheochromocytoma Cells. Endocrine Soc. Abst., San Francisco, 2002.

Walters, R.J., Rosenbaum, C, Willenberg, H.S., Ehrhart-Bornstein, M, Muller, H-W, Siebler, M., Scherbaum, W.A., Bornstein, S.R. Insulin Induces Increased Expression and Focal Clustering of GABA-A Receptors in PC12 Cells. Endocrine Soc. Abst. (platform), San Francisco, 2002.

R.J. Walters. A fourth strategy to contain the threat of HIV and AIDS? *New York Science Times Article February 1996.*

Barber, M., Collier, J., Walters, R.J. The HiPaCC Diet, *2007.*

Walters, R.J. Mind Bomb: The predictive power of reason. *2014*

www.ingramcontent.com/pod-product-compliance
Lightning Source LLC
Chambersburg PA
CBHW082324220526
45470CB00008B/2392